S0-BRA-837

Mathematics for Elementary School Teachers: Explorations

FOURTH EDITION

Mathematics for Elementary School Teachers: Explorations

Tom Bassarear
Keene State College

HOUGHTON MIFFLIN COMPANY   Boston   New York

Publisher: *Richard Stratton*
Senior Sponsoring Editor: *Lynn Cox*
Senior Marketing Manager: *Jennifer Jones*
Marketing Associate: *Mary Legere*
Senior Development Editor: *Maria Morelli*
Editorial Assistant: *Laura Ricci*
Associate Project Editor: *Kristen Truncellito*
Art and Design Manager: *Gary Crespo*
Cover Design Manager: *Anne S. Katzeff*
Senior Photo Editor: *Jennifer Meyer Dare*
Composition Buyer: *Chuck Dutton*

Cover photograph: © *Purestock/Getty Images*

Copyright © 2008 by Houghton Mifflin Company. All rights reserved.

Houghton Mifflin Company hereby grants you permission to reproduce the Houghton Mifflin material contained in this work in classroom quantities, solely for use with the accompanying Houghton Mifflin textbook. All reproductions must include the Houghton Mifflin copyright notice, and no fee may be collected except to cover the cost of duplication. If you wish to make any other use of this material, including reproducing or transmitting the material or portions thereof in any form or by any electronic or mechanical means including any information storage or retrieval system, you must obtain prior written permission from Houghton Mifflin Company, unless such use is expressly permitted by federal copyright law. If you wish to reproduce material acknowledging a rights holder other than Houghton Mifflin Company, you must obtain permission from the rights holder. Address inquiries to College Permissions, Houghton Mifflin Company, 222 Berkeley Street, Boston, MA 02116-3764.

Printed in the U.S.A.

Library of Congress Control Number: 2006921991

ISBN-10: 0-618-76837-8
ISBN-13: 978-0-618-76837-0

4 5 6 7 8 9 — HS — 11 10 09 08

Contents

Copyright © Houghton Mifflin Company. All rights reserved.

Copyright © Houghton Mifflin Company. All rights reserved.

6   Proportional Reasoning 131

7   Uncertainty: Data and Chance 143

Copyright © Houghton Mifflin Company. All rights reserved.

8 Geometry as Shape   175

Copyright © Houghton Mifflin Company. All rights reserved.

9 Geometry as Transforming Shapes   231

10 Geometry as Measurement   293

Copyright © Houghton Mifflin Company. All rights reserved.

Index

Cutouts

Base 10 Graph Paper

Other Base Graph Paper

Other Base Graph Paper

Other Base Graph Paper

Other Base Graph Paper

Geoboard Dot Paper

Isometric Dot Paper

Polyomino Grid Paper

Polyomino Grid Paper

Polyomino Grid Paper

Polyomino Grid Paper

Tangram Template

Regular Polygons

Exploring the Area of a Circle

Copyright © Houghton Mifflin Company. All rights reserved.

Preface

Having a textbook contain two different volumes created the dilemma of where to put the preface. We resolved this dilemma by having two prefaces. This one, to the Explorations, sets the overall theme of the course and explains how the two volumes work with each other. The preface to the Text more specifically describes the goals of the course and the textbook features. For semantic reasons, this preface is addressed to the student and the Text Preface is addressed to the instructor. However, I have written each with the belief that student and instructor will find both prefaces useful.

I have been teaching the Mathematics for Elementary School Teachers course for twenty years, and in that time have learned as much from students like you as they have learned from me. The Explorations and the Text reflect the most important things we have taught each other: that building an understanding of mathematics is an active, exploratory process and, ultimately, a rewarding, pleasurable one. Many students have told me and other instructors that this is the most readable and interesting mathematics textbook they have ever read. This is exciting to hear, because I thoroughly enjoy mathematics and hope others will too. At the same time, I know that far from loving mathematics, many people are actually afraid of it. If this book is successful, you will come to believe that math can be enjoyable and interesting (if you don't already feel that way); that mathematics is more than just numbers and formulas and is an important part of the curriculum, and that mathematical thinking can be done by "regular" people.

What is it we want you to learn?

At its heart, the purpose of this course is to revisit the content of pre-K–8 mathematics so that you can build an integrated understanding of mathematical concepts and procedures. From one perspective, it is useful to talk about content knowledge (e.g., different meanings of subtraction, why we need a common denominator when adding fractions, and classification of geometric figures) and process knowledge (e.g., problem-solving and being able to communicate mathematical ideas and solutions). Let me give some examples of the kinds of content knowledge you can expect to learn in this book.

Content Knowledge Virtually all adults can do the division problem at the right. If asked to describe the procedure, most descriptions would sound something like this: "4 'gazinta' (goes into) 9 two times, put the 2 above the 9, 4 times 2 is 8, 9 − 8 is 1, bring down the 2, 4 gazinta 12 three times with no remainder, put the 3 above the 2, the answer is 23." However, few adults can explain mathematically what "goes into" and "bring down" mean, other than by saying "that's how you do it." However, once you realize that one (of several) meanings of division is that an amount is to be distributed equally into groups and once you realize that 92 can be represented as 9 tens and 2 ones, then you can explain why long division works. Being able to explain *why* gives one what we call mathematical power, which means, among other things, that you can apply this knowledge to solve problems that are not just like the ones in the book. You can solve "real-life" problems, which is one of the most important goals of school mathematics.

Similarly, most elementary teachers recall that an approximation of π (pi) is 3.14 and they may remember some formulas: $C = \pi d$, $C = 2\pi r$, $A = \pi r^2$, but they can't explain what π means or why these formulas work. Developing a conceptual understanding of

$$\begin{array}{r} 23 \\ 4\overline{)92} \\ \underline{8} \\ 12 \\ \underline{12} \\ \end{array}$$

Copyright © Houghton Mifflin Company. All rights reserved.

π is not as esoteric as many people think. Let me illustrate. Look at the circle at the right and answer the following question: Imagine we had several flexible rulers that were the same length as the diameter. If you wrapped those rulers around the circle, how many rulers would it take to wrap around the entire circle? If you actually do this, you find that it will take a little more than three rulers. Thus, one conception of π is that the length of the circumference of a circle is always a bit more than 3 times the length of the diameter. With this conceptual understanding, the formula C = πd "makes sense." Ensuring that math makes sense is another key goal of mathematics education.

I have just outlined two of many explorations and investigations that you will do over the course of this book. What so many of my students have discovered is that if they have a chance to work with mathematical concepts in an active, exploratory manner, they can make sense of elementary mathematics, which means that they will be much more effective teachers. Much of the impetus for teaching to foster this style of learning comes from the National Council of Teachers of Mathematics (NCTM), which has published three sets of Standards in the past twenty years and ushered in a new reform movement in mathematics education. I find that many students who enter the course with negative feelings toward mathematics view the NCTM Standards very positively. Many tell me, "I wish this is what my mathematics courses had been like."

NCTM   The first NCTM Standards document, *Curriculum and Evaluation Standards*, was published in 1989. It sets forth a vision of why mathematics is important for all citizens to know and describes the mathematical knowledge one should develop by the end of high school. Chapter 1 of the Text will explore the Curriculum Standards in more detail. The *Professional Standards for Teaching Mathematics* will be discussed later, and you should expect to examine the *Assessment Standards for School Mathematics* in your methods course. Since one of my goals is that you will use NCTM to frame your own teaching, I will discuss my vision of using NCTM Standards as much as possible. At appropriate times in the text, I will cite passages from the NCTM documents.

In 2000, the NCTM finalized a new document called *Principles and Standards for School Mathematics*, which represents an update and refinement of the 1989 document. A summary of the ten standards can be found in Appendix A of the main text, and the full document can be found at the NCTM website: nctm.org.

What does it mean to learn?

As you will discover in Chapter 1, your attitudes and beliefs have a lot to do with how you learn mathematics. My beliefs about what it means to "know" mathematics have led me to create a very different kind of textbook, and it is important for me to describe my sense of what it means to learn. Let me contrast some traditional beliefs about mathematical learning (which I disagree with) with this book's approach, and then describe how this book is structured.

Active versus Passive Understanding   Many students believe that mathematical understanding is either-or; e.g., you either know fractions or you don't. In actuality, mathematical understanding is very much like other kinds of knowledge—you "sort of" know some things, you know other things "pretty well," and you know some things very well. The belief in either-or leads many students to focus on trying to get answers instead of trying to make sense of problems and situations. However, when learning is seen as an "it" that the students are supposed to "get," the student's role is seen more passively. Our language betrays this bias—the teacher "covers the material" and the students "absorb." Steven Leinwand, from the Connecticut State Department of Education, once told a joke at a mathematics teachers' convention about the Martian who came to visit American schools. In her report to the Martians, she said, "Teachers are people who work really hard and students are people who watch teachers working really hard!" In

Copyright © Houghton Mifflin Company. All rights reserved.

a similar vein, the NCTM has a button that reads: "Mathematics is not a spectator sport." Much of the joy of mathematics is examining a situation or problem and trying to understand it. My own experience with elementary school children and my own two children, Emily and Josh, has convinced me that young children naturally seek to make sense of the world that they live in, and that for a variety of reasons many people slowly lose that curiosity over time.

Worthwhile Mathematical Tasks versus Oversimplified Problems Another common belief about learning mathematics is that we should make the initial problems as simple and straightforward as possible. I call this approach the Lysol approach—we clean up the problems to reduce the complexity (we use routine problems and one-step problems), we try to make the problems unambiguous, and we get rid of extraneous information (we make sure to use every number in the problems). However, one of the biggest drawbacks of this approach is that most of the problems that people solve outside of classrooms are complex and ambiguous, and part of the problem is to determine what information is relevant. A famous Sufi story nicely illustrates the difference between the two approaches. Nasruddin came home one night and found a friend outside on his hands and knees looking in the dirt. When Nasruddin asked, "What's happening?" his friend replied, "I dropped my keys." Nasruddin asked, "Where did you drop them?" His friend pointed: "Over there." Puzzled, Nasruddin asked, "Then why are we looking here?" The response: "Oh, the light is much clearer here." Although the light is much clearer when the problems are unambiguous, routine, and one-step, that is just not how students learn to think mathematically. One consequence of these richer problems (NCTM uses the term "worthwhile mathematical tasks") is that most students report that the problems are much more interesting than what they generally find in texts.

Owning versus Renting Another way of describing these two very different beliefs about learning mathematics that many of my students have found illuminating is to talk of the difference between owning and renting knowledge. Many students report that one year after finishing a mathematics course, they have forgotten most of what they learned—i.e., if they retook the final exam, they would fail. This means that they rented the knowledge they learned: they kept it long enough for the tests, and then it was gone. However, this need not be so. If you are an active participant in the learning process and if the instructional strategies of the professor fit with how you learn, then you will own most of what you learn. When you take your methods course in a year or so, you will still remember the important ideas from this course. It is such a wonderful feeling to know that you got more than just 3 or 4 credits for the 100+ hours you spent over the course of the semester. If you plan to be an excellent elementary teacher (and I expect you all do), then you need to own what you learn.

An Integrated Approach If we want students to retain the knowledge, then how they learn it has to be different than it has been in traditional classrooms. One of my favorite words, and one you will find in the NCTM Standards, is *grapple*. I have found that if students grapple with problems and situations and try to make sense of them, they are more likely to retain what they learn in the process. I believe that a central part of the teacher's job is to select worthwhile tasks (Professional Standard 1) and to develop an environment that invites all students to learn and that honors differences in how they learn (Professional Standard 5). A classroom consistent with the NCTM vision does not look like a traditional classroom in which the teacher mostly lectures and demonstrates, and students generally take notes, ask questions, and answer questions the teacher asks. Rather, the classroom looks like an ongoing dialogue: the teacher presents a problem, possibly a brief lecture, and then the teacher facilitates the discussion around that question or situation (Professional Standards 2, 3, and 4), during which students naturally expect to make guesses (predictions and hypotheses) and try to explain their thinking and justify their hypotheses. Similarly, I believe that the textbook for this classroom must

Copyright © Houghton Mifflin Company. All rights reserved.

be different from a traditional book in which the important concepts and formulas are highlighted and in which the problems at the end of the chapter are generally pretty much like the examples.

In the introduction to NCTM's *Professional Standards for Teaching Mathematics*, the authors summarize Major Shifts in mathematics classrooms that have been called for by the NCTM. This passage nicely summarizes much of what I have just described.

We need to shift:

- toward classrooms as mathematical communities—away from classrooms as simply collections of individuals;
- toward logic and mathematical evidence as verification—away from the teacher as the sole authority for right answers;
- toward mathematical reasoning—away from merely memorizing procedures;
- toward conjecturing, inventing, and problem-solving—away from an emphasis on mechanistic answer finding;
- toward connecting mathematics, its ideas, and its applications—away from treating mathematics as a body of isolated concepts and procedures.

If we can convince our students that fundamentally mathematics is a sense-making enterprise, then we change not only how much they work but also *how* they work. That is, if they see the relevance of the problems and concepts and develop confidence, they work harder; if they see the goal as sense making versus "getting it," they work differently. With this brief overview of what I mean by learning, let me explain how the two volumes work together.

Role of Explorations and Text

As just mentioned, if we believe that people learn better by grappling with richer problems (Worthwhile Mathematical Tasks) as opposed to being shown how to do simpler problems, then this creates a whole host of changes in what the classroom looks like, what the instructional materials look like, and how students' knowledge is assessed. Basically, these Explorations have been designed to have you grapple with important mathematical ideas. I have posed questions and tasks that call on you to discover fundamental mathematical concepts and structures, for example, that percents can be seen as proportions. In the Explorations, you grapple with new ideas and concepts in a hands-on environment. The Text then serves as a resource: a place where the formal definitions and structures are laid out (to be used generally after the Explorations), a place to refine your understanding, a place to self-assess, and of course, a source of homework problems. The Explorations in this volume are generally open-ended and often have more than one valid answer, whereas the Investigations (in the Text) are generally more focused in scope and, though they generally have one correct answer, the discussions following the Investigations show that they always lend themselves to different ways to arrive at that answer. I generally begin each new chapter with an Exploration that, in learning theory terms, enables and encourages students to construct a scaffolding of major mathematical concepts. I then use the Text as a resource where the students can see the concepts articulated in an organized way.

Looking back is a habit of mind that I want to encourage. A famous mathematics teacher proverb is "you're not finished when you have an answer; you're finished when you have an answer that makes sense to you." I model this notion explicitly in Chapter 1 in the Explorations and in the Text and you'll see it often in both volumes. The idea of reflection is new to many students, but getting into the habit of looking back is a great way to take ownership of this content. With this in mind, I hope you're looking forward to exploring elementary mathematics.

Copyright © Houghton Mifflin Company. All rights reserved.

b. Predict the 10th row and explain your prediction. Then color the 10th row, and write down any new ideas if your prediction was wrong.

c. Continue this process of predicting, justifying, and checking through the 16th row.

d. Predict what spaces would be colored in the 17th row if your worksheet went that far. Justify your thinking.

e. How many more rows can you confidently predict? Describe your reasoning.

f. Write any new patterns you see now.

5. Repeat the same process with other multiples that your instructor assigns.

6. Examine the Visual Patterns Resource Page on page 19.

a. What patterns do you see that appear in some or all tables?

b. If a pattern appears in some, but not all, tables, what kind of relationship do the numbers at the bottom of the table have to each other?

7. We can do a similar process by coloring remainders when divided by 3. Use three highlighter pens that are different colors. For each number in the triangle, divide it by 3 and see what the remainder is.

- If there is no remainder, color that cell one color.
- If the remainder is 1, color it using another color.
- If the remainder is 2, color it using still another color.

As with the previous coloring problems, look for patterns. Try to predict the colors of the cells in the next row. After coloring 8 rows, stop.

a. What patterns do you see?

b. Predict the colors of the 9th row. Color in the 9th row. If you were right, great. If you were wrong, write down what you see now that you didn't see before.

c. Predict the colors of the 10th row, explain your reasoning, color the 10th row, and write down any new ideas if your prediction was wrong.

d. Continue through the 16th row. Write any new patterns you see now.

e. Predict the colors of the 17th row.

f. How many more rows can you confidently predict? Describe your reasoning.

PROBLEM 2: How many handshakes?

Determine how many students are in your class, including yourself. If each student shakes hands with every student, how many handshakes will there be?

1. Work on this problem alone for a few minutes. Can you apply ideas discussed in the preface to find patterns in this problem? Can you use what you see to help you plan a solution?

2. Discuss your ideas. Describe new ideas that you like that arose from the discussion.

3. Now solve the problem. Refer to "4 Steps for Solving Problems" as needed.

4. After the class discussion of the problem, select a solution that was different from the one you used. Explain that way of doing the problem, as if to a student who was not in class today.

5. The next step here is to generalize our work. That is, if there were n students in the class, how many handshakes would there be?

a. Find the answer and show your work.

b. Explain where the equation comes from. That is, here we are not simply interested in knowing the answer but where it comes from. In this respect, we are beginning a semester-long discussion of the notion of reasoning and proof which is discussed in Chapter 1.

Copyright © Houghton Mifflin Company. All rights reserved.

6. On Tape 17 in the *Teaching Math: A Video Library, K–4* series, a fourth-grade teacher poses the following question to her class: If all 24 children in the class exchange Valentine's Day cards, how many cards will be needed? Do you think the answer to this question will be the same as the number of handshakes there will be for 24 people? Why or why not?

7. If you see connections between this problem and parts of Problem 1, describe them.

PROBLEM 3: How many ways to make money and stamps

1. How many different ways can you make 25¢ with coins?
    Think first about your method. Randomly thinking of different combinations will not help you to grow as a mathematics student. Thus, think about how you might be systematic (there are different ways, as you will soon see) and think about how your representation of the problem might help you to find all of the combinations.

    a. Now find and show all the combinations in some kind of table.
    b. Describe at least three patterns you see in your table.

2. How many different ways can you make 50¢ with coins?
    A key goal in this problem is to continue with your development in being systematic, looking at representations that will be more useful, *and* to think about how you can use the work from the 25¢ problem to answer this question.

    a. Find and show all the combinations in some kind of table.
    b. Describe at least three patterns you see in your table.
    c. Describe how you used your work from the step in part (a) to solve this problem.

3. Here are several coin problems that have a different flavor. In each case, the main goal is not simply to get the answer, but to think carefully about tools that you can use other than just random guess–test–revise. Thus, in each case, show your answer, show your solution path, and briefly summarize your method.

    a. How can you have 5 coins that add up to 55 cents? Is there another solution? Can you *prove* that there is or isn't?
    b. How can you have 9 coins that add up to 60 cents?
    c. How can you have 10 coins that add up to 60 cents?
    d. How can you have 21 dimes, nickels, and pennies that add up to one dollar? This was a Math Forum problem, for which student work can be seen at **http://mathforum.org/library/drmath/view/59182.html**
    e. Make up your own problem.

4. Stamps
    As with the other problems, first think about how you can go about generating the possibilities systematically and then how you want to represent your work. In each problem, find and show all the combinations in some kind of table, and describe at least three patterns you see in your table.

    a. How many different ways can you make 37¢ postage with 10¢, 3¢, and 2¢ stamps?
    b. How many different ways can you make 40¢ with 8¢, 5¢, and 2¢ stamps?

5. Make up your own problem.

Copyright © Houghton Mifflin Company. All rights reserved.

together. Continue this process until the sum is a palindrome. The example below shows why 68 is a 3-step palindrome.

$$
\begin{array}{r} 68 \\ + 86 \\ \hline 154 \end{array}
\qquad
\begin{array}{r} 154 \\ +451 \\ \hline 605 \end{array}
\qquad
\begin{array}{r} 605 \\ + 506 \\ \hline 1111 \end{array}
$$

Explore all two-digit numbers from 11 to 99 and determine which numbers are one-step palindromes, two-step palindromes, three-step palindromes, etc. As you go through the numbers from 11 to 99, look for patterns or observations that can shorten your work. For example, I know this is a 1-step palindrome because . . . , I know this is a two-step palindrome because. . . .

1. Make a table like the one below that lists the already palindromes, one-step palindromes, two-step, etc.

Already	1-step	2-step	3-step	4-step	5-step
11	14	48			

2. Describe the characteristics of members of each family. In some cases, all of the members of that family have the same characteristics. In some cases, you will have to say—these x-step palindromes have this characteristic and these y-step palindromes have this characteristic.

3. Make a table for the numbers 11–99 with 9 rows; the beginning number of the rows are 11, 21, 31, 41, etc. Color the already palindromes with one color; color the 1-step palindromes with another color; color the 2-step palindromes with another color, etc. What do you see?

Copyright © Houghton Mifflin Company. All rights reserved.

EXPLORATION 1.5 Master Mind

We will play a variation of a game called Master Mind that has entertained people for many years. I have seen this game in many elementary school classrooms. Teachers have used it to develop logical thinking, and it is a popular choice in classrooms where students are given a chance to spend some time on any activity they choose.

Directions for playing the game:

- One group will think of a four-digit number.

- The second group's goal is to guess the number.

- After each guess, the first group will tell how many digits are correct *and* how many digits are in the correct place. For example, if the number is 1234 and the guess is 2468, the feedback will be two correct digits and none in the correct place.

- Your instructor may limit the game at first, requiring, for example, that players use only the digits 1, 2, 3, 4, and 5 with no repetitions; you can't have 3333, for example.

1. Play the game several times.

2. Answer the following questions within the parameters of your game.

   a. If you are in the first group, what number or sets of numbers are you more likely to select, assuming you want to make the number harder to solve? Why?

   b. If you are in the second group, what number would be a good first guess? Why?

   c. Let's say you know some students in another class who haven't played the game yet. What hints would you give them?

Copyright © Houghton Mifflin Company. All rights reserved.

EXPLORATION 1.6 Magic Squares

Magic squares have fascinated human beings for many thousands of years. The oldest recorded magic square, the Lo Shu magic square, dates to 2200 B.C. and is supposed to have been marked on the back of a divine tortoise that appeared before Emperor Yu when he was standing on the bank of the Yellow River. In the Middle Ages, many people considered magic squares to be able to protect them against illness! Even in the twenty-first century, people in many countries still use magic squares as amulets.

We will spend some time exploring magic squares because they reveal some amazing patterns. As a teacher, you will find that many of your students love working with magic squares and other magic figures.

PART 1: Describing magic squares and finding patterns

1. Let us begin with a simple magic square. The definition of a magic square is that the sum of any row, any column, and each diagonal is the same, in this case 15. However, there is much more. Look at the square and write down anything that you observe in this square—relationships between numbers in rows or columns or diagonals, patterns in how the numbers are arranged, even/odd, etc.

8	1	6
3	5	7
4	9	2

2. Compare your observations with others in your group. Add to your list new observations and patterns that you heard from other members.

PART 2: Patterns in all 3 × 3 magic squares

1. You will find eight different 3 × 3 magic squares on page 20. Look at these magic squares carefully and then write down

   a. Patterns that are found in all of the magic squares
   b. Patterns that are found in some of the magic squares

2. Compare your observations with others in your group. Add to your list new observations and patterns that you heard from other members.

3. Now that you have seen nine different magic squares, do you see patterns that would enable you to make your own magic square? If so, write down your thoughts about how to make other 3 × 3 magic squares. If not, read on.

PART 3: Using algebra to describe magic squares

One of the objectives of this course is to help students realize that algebraic thinking does not begin in high school and to appreciate how simple algebraic notation can make life simpler! This part is designed to help students see this.

1. As you may have already discovered, the middle number of the magic square is a key number. What if we called the middle number m? Reflect on the patterns you observed in the magic squares. Can you represent the other numbers in the magic square in terms of m? Describe your present thinking and work before moving on.

Copyright © Houghton Mifflin Company. All rights reserved.

2. One key to solving this problem is to realize how many more variables are needed. One representation requires two more variables. Let us call them x and y. In other words, using m, x, and y, we can represent the value of each cell in the magic square. If you were stuck in question 1, see if you can develop a representation for all 3×3 magic squares using these three variables.

3. At this point, you have a solution generated either from your group or from the whole-class discussion. Think about your work and discussions and then read the following quotation: "Mathematics is often considered a difficult and mysterious science, because of the numerous symbols which it employs. . . . [T]he technical terms of any profession or trade are incomprehensible to those who have never been trained to use them. But this is not because they are difficult in themselves. On the contrary they have invariably been introduced to make things easy. So in mathematics, granted that we are giving any serious attention to mathematical ideas, the symbolism is invariably an immense simplification."[1] Does this experience change your attitude toward algebra? Does it help you to see the use of symbols in a new light?

PART 4: Further explorations

1. How many different 3×3 magic squares can you make starting with these two numbers?

2. Below are questions about four possible transformations of a 3×3 magic square. In each case, write your initial guess and your justification before you test your guess. Then test your guess. If you were correct, refine your justification if needed. If you were wrong, look for a flaw or incompleteness in your reasoning. If you were wrong, can you now justify the correct answer?

   a. What if you doubled each number in a magic square? Would it still be a magic square?

   b. What if you added the same number to each number in a magic square? Would it still be a magic square?

   c. What if you multiplied each number in a magic square by 3 and then subtracted 2 from that number? Would it still be a magic square?

   d. What if you squared each number in a magic square? Would it still be a magic square?

3. Look back on your work from different 3×3 magic squares, and answer the following questions:

   a. Is there a relationship between the magic sum and whether the number in the center is even or odd?

   b. Divide the set of magic squares into two subsets: those in which the nine numbers are consecutive numbers (such as 10–18) and those in which the nine numbers are not consecutive numbers. Are there any other differences between these two sets of magic squares?

[1]*Introduction to Mathematics* (New York, 1911), pp. 59–69, cited in Robert Moritz, *On Mathematics* (New York: Dover Publications, 1914), p. 199, quoted: A. N. Whitehead.

Copyright © Houghton Mifflin Company. All rights reserved.

EXPLORATION 1.7 Magic Triangle Puzzles

Many children enjoy problems like the ones in this exploration, and the problems contain great opportunities to develop mathematically. In addition to providing computation practice, they also help children develop more powerful problem-solving strategies and develop their reasoning abilities.

PART 1: Magic triangles

The task is to determine the numbers that belong in the circles so that the sum of the numbers in any two circles equals the number between them. The first problem is worked out for you below.

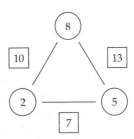

1. Now, solve the puzzles below. As you play with these problems, look for patterns that might unlock the problem so you don't have to do random guess-test-revise every time:

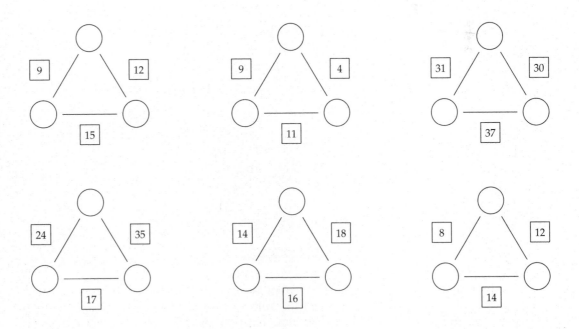

2. Once you have a method that will solve any triangle quickly, write your directions for your method and give them to someone not in this class. Ask them to solve the triangle probelms just on your directions. If they get stuck, ask them what part of your directions is confusing. Straighten out that confusion and revise your directions.

Copyright © Houghton Mifflin Company. All rights reserved.

3. Give the revised directions to another person. How did they do?

4. What did you learn about writing clear directions?

5. Make up some of your own probelms.

PART 2: A different kind of magic triangle puzzle.

This exploration is adapted from "Building a Community of Mathematicians" in the April 2003 issue of *Teaching Children Mathematics.**

1. Using the numbers 1 through 6, place numbers in each circle so that when you add the three numbers that make up each side of the triangle, you get the same sum in all three cases. How many different solutions can you find?

2. Look again at all the solutions that your class found. What do you see that is interesting? What patterns do you see?

3. Now draw line segments inside the puzzle so that the numbers in the middle of the sides of the triangles are connected. You can now break each puzzle into three small triangles, each comprising three numbers. What patterns or relationships do you see in these triangles?

*Reprinted with permission from *Teaching Children Mathematics,* copyright © 2003 by the National Council of Teachers of Mathematics.

Copyright © Houghton Mifflin Company. All rights reserved.

Pascal's Triangle

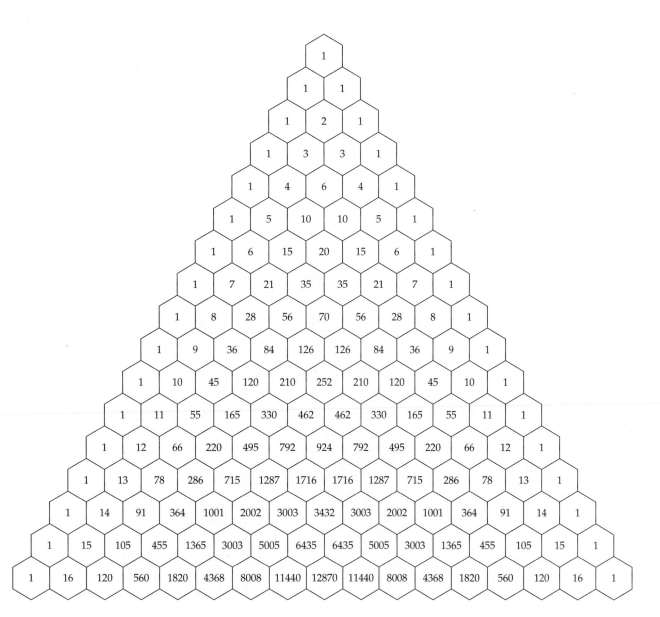

Copyright © Houghton Mifflin Company. All rights reserved.

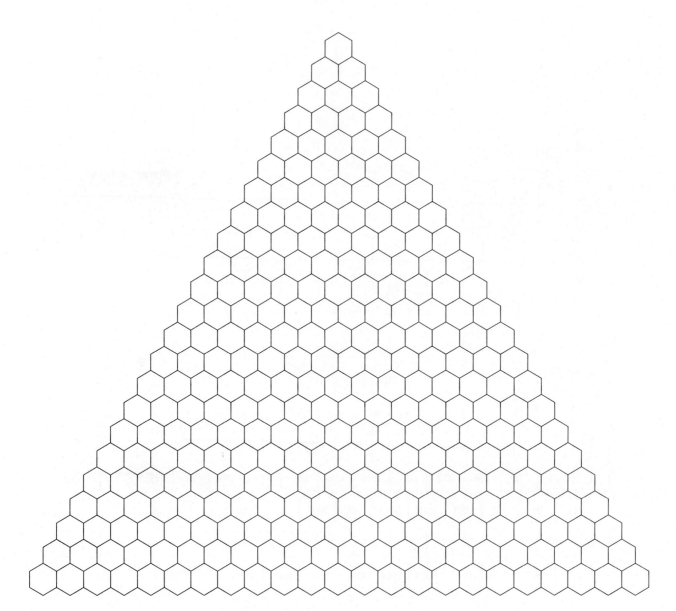

Copyright © Houghton Mifflin Company. All rights reserved.

Visual Patterns Resource Page*

Multiples of 2

Multiples of 3

Multiples of 4

Multiples of 5

Multiples of 6

Multiples of 7

Multiples of 8

Multiples of 9

Multiples of 10

*From *Pascal's Triangle* by Dale Seymour. Copyright © 1986 by Pearson Education, Inc., publishing as Dale Seymour Publications, an imprint of Pearson Learning Group. Used by permission.

Copyright © Houghton Mifflin Company. All rights reserved.

3 × 3 Magic Squares and Templates for EXPLORATION 1.6

18	33	15
19	22	25
29	11	26

7	17	3
5	9	13
15	1	11

6	10	11
14	9	4
7	8	12

7	26	6
12	13	14
20	0	19

14	28	12
16	18	20
24	8	22

9	26	7
12	14	16
21	2	19

10	21	8
11	13	15
18	5	16

6	13	8
11	9	7
10	5	12

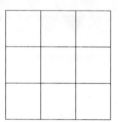
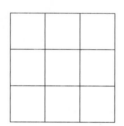
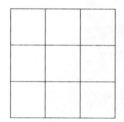
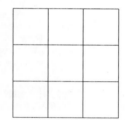
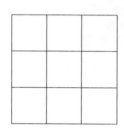
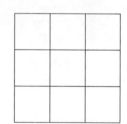

Copyright © Houghton Mifflin Company. All rights reserved.

3. How many dots are in the nth triangular number? Explain your prediction and how you came to your conclusion.

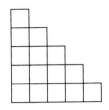

 If you have trouble with this question, look back in Chapter 1 to find any problems that connect to this question. Or represent the numbers on graph paper with squares instead of dots (see the figure at the right) and connect the question to what part of a complete rectangle has been made.

Relationships Between Figurate Numbers

4. *Interpreting a new notion* The following formula expresses *one* way in which square and triangular numbers are related:

$$S_n = T_{n-1} + T_n$$

   a. Translate this equation into English.
   b. Exchange your translation with a partner and discuss each of your translations until you both feel satisfied that both phrasings work.
   c. Justify this relationship—that is, explain why it is true. Write your first draft of a justification.

5. *Translating words into notation*

   a. Translate this sentence into notation: The square of any odd number is one more than eight times a specific triangular number.
   b. Exchange your notation with a partner and discuss each person's work.

Other Figurate Numbers

6. Look at the pentagonal numbers below.

 Describe patterns that you observe and how you can predict the 5th, 6th, and nth pentagonal numbers.

7. Look at the hexagonal numbers below.

 Describe patterns that you observe and how you can predict the 5th, 6th, and nth hexagonal numbers.

Copyright © Houghton Mifflin Company. All rights reserved.

8. Make a table like the one shown below. Write in your expressions for the value of the *n*th triangular number and the *n*th square number. Determine the value of the *n*th pentagonal number and the *n*th hexagonal number. Show your work.

Type of number	Value of the *n*th term
Triangular	
Square	
Pentagonal	
Hexagonal	
*n*th figurate number	

9. Can you predict the value of the *n*th number for *any* figurate-number family, such as the *n*th octagonal number?

PART 2: Squares around the border

The following problem is a popular growing pattern problem in elementary schools because it is so rich. The February 1997 issue of *Teaching Children Mathematics* has a whole article describing how this problem can be adapted for grades K–2, for grades 3–4, and for grades 5–6. I encourage you to look it up in your library. I have also seen variations of this problem at several conferences and in other articles. With young children, the problem is often presented with blue and white tiles (that you can get at a tile store). The blue tiles (in the center) represent the water in the pool, and the white tiles represent the border around the pool.

1. Below you can see the first five figures in this growth pattern. Count the number of squares in the border of each figure, and then, from the patterns you see and the observations you make, determine the number of squares around the border of the *n*th figure. As in Part 1, we will break the problem into a series of smaller steps.

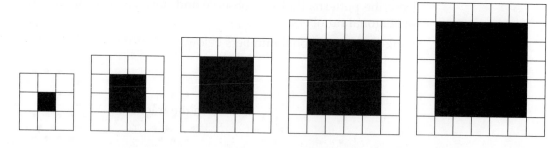

a. How many squares are there in the border of the first figure? of the second figure? of the third figure? of the fourth figure? What patterns do you see, in the figure and/or from how you counted the number of squares? Draw additional figures if you feel that would help.

b. Your instructor will now have you share your observations with your partners or the whole class.

Copyright © Houghton Mifflin Company. All rights reserved.

c. Working at first on your own and then with your partners, try to make use of observations and patterns to determine the number of squares in the border of the nth figure.

2. One of the themes of the book is the notion of multiple representations. Often, we see new aspects of a situation when we examine it from another perspective. We will do that here—the other perspective is to see what graphs tell us about the relationships among the figures, the number of squares around the border, and the number of squares in the middle.

a. Make a table like the one below. Fill it out and then make two graphs. In the first graph, the independent variable will be the number of the figure (1st, 2nd, 3rd, etc.), and the dependent variable will be the number of squares around the border of that figure. In the second graph, the independent variable will again be the number of the figure, but the dependent variable will be the number of squares in the middle of that figure.

b. Record any observations you make from the two graphs.

c. How would you describe, in words, the growth in the number of squares around the border and the growth in the number of squares in the middle?

Figure number	Number of squares around the border	Number of squares in the middle	Notes on how you determined this
1			
2			
3			
4			
5			
6			
7			

Extensions

As you are finding, many rich problems have more to offer even after you have found the answer. They are extendible! Let us examine three extensions of the squares-around-the-border problem.

3. a. Determine the fraction of the total area that is contained by the "pool" (inner shaded region) for the first five figures.
    b. Describe the changes you see in the fraction.
    c. What fraction will be pool in the nth figure?
    d. Make a graph of this situation: The independent variable is the number of the figure, and the dependent variable is the fraction of the area that is pool. Describe this function in words.

Copyright © Houghton Mifflin Company. All rights reserved.

4. The growth pattern shown below adds one more layer to this problem. We have the white outer border and a shaded inner border. How many squares would make up the nth outer border (white squares)?

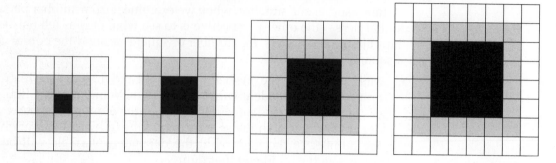

5. This extension comes from realizing that we don't have to have a square "pool" and a square "border." How many squares are in the border of the nth figure for the growth pattern shown below?

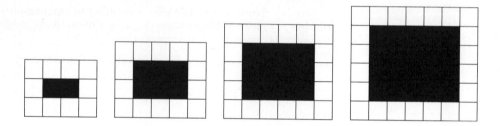

Copyright © Houghton Mifflin Company. All rights reserved.

SECTION **3.1** Exploring Addition

Can you remember learning to add and subtract in first and second grade? As adults you probably don't think twice about adding and subtracting, but these operations are often difficult for many youngsters, especially when regrouping is involved. One way to help you better understand what these operations mean is to have you do problems in a different setting.

In Chapter 2 you put yourself in the role of a member of the Alphabitian tribe (Exploration 2.8), and in that role you created a numeration system. You will be asked to remain Alphabitians for a while longer and to add and subtract in that system.

EXPLORATION 3.2 Mental Addition

1. Do in your head the eight computations below. Briefly note the strategies that you used, and try to give names to them.

    One mental tool that all students have is being able to visualize the standard algorithm in their heads. For example, for part (a), you could say, "9 + 7 = 16, carry the 1, then 5 + 3 = 8 + the carried 1 makes 9; the answer is 96." However, because you already own that method, I ask you not to use it here but to try others instead.

2. Share your strategies in your small group. Note any strategies that you heard that you did not use but would like to use.

3. In your group, select two or three strategies to describe to the class. Make up a name for each strategy.

4. After hearing the class presentations, write down the strategies that you like best.

Sum	What you did	Name
a. 39 + 57		
b. 27 + 58		
c. 78 + 25		
d. 46 + 19		
e. 625 + 147		
f. 588 + 225		
g. 790 + 234		
h. 8734 + 467		

Copyright © Houghton Mifflin Company. All rights reserved.

EXPLORATION 3.3   Addition: Children's Algorithms and Alternative Algorithms

PART 1: Children's algorithms

For each of the operations, we will examine some algorithms that children invent and some algorithms that have been and/or continue to be popular in other parts of the world.

Below are descriptions of several solution paths for the problem 48 + 26—paths that children commonly invent when they are not shown how to add but, rather, are provided with rich questions to develop their understanding of place value and then are asked to figure out these problems on their own. In each case,

a. Make up and solve additional problems until you understand how the algorithm works.

b. Determine whether that solution path will work for larger numbers: for 385 + 562, for 476 + 508, for 3245 + 683.

c. Write your first draft of why the algorithm works.

1. Add up.
   The child says: 48 + 20 = 68; 68 + 6 = 74

2. Add each place, and then combine.
   The child says: 40 + 20 is 60, 8 + 6 = 14, 60 + 14 = 74

3. Compensate.
   The child says: take 2 from 26 and give it to 48; 50 + 24 = 74

PART 2: Alternative algorithms

Below are several algorithms that have enjoyed popularity in different countries at different times. In each case,

a. Make up and solve more problems until you understand how the algorithm works.

b. Determine whether it will work for larger numbers: for 385 + 562, for 476 + 508, for 3245 + 683.

c. Write your first draft of why the algorithm works.

d. Compare the algorithm to the standard addition algorithm in the text with respect to the five criteria Hyman Bass described, which are given on page 162 of the text.

Cross out

Add from left to right. When you see the needed regrouping in the place to the right, cross out and increase the current place by 1.

$$
\begin{array}{r}
358 \\
+\ 574 \\
\hline
8\!\!\!/22 \\
93
\end{array}
$$

3 + 5 = 8; look to the right—because 5 + 7 > 9, put a 9 in the hundreds place.

5 + 7 = 12; look to the right—because 8 + 4 > 9, put a 3 in the tens place.

8 + 4 = 12; put a 2 in the ones place.

Copyright © Houghton Mifflin Company. All rights reserved.

The "Scratch" or "Adding Up" Algorithm

At the right, you can see how 564 + 378 is computed with the "scratch" or "adding up" algorithm. You start with the ones place and find the sum of each place. If the sum is less than 10, you simply record the sum at the top. If the sum is 10 or greater, you put the digit from the ones place of the sum above and then cross out the number in the place to the left and increase it by 1. This algorithm is quite probably the origin of the phrase *adding up*.

$$
\begin{array}{r}
94 \\
8\!\!\!/32 \\
564 \\[4pt]
378
\end{array}
$$

The Partial Sums Algorithm

The partial sums algorithm was developed in India over 1000 years ago. It works like this:

$$
\begin{array}{r}
369 \\
478 \\
\hline
17 \\
13 \\
7 \\
\hline
847
\end{array}
$$

The Lattice Algorithm

The lattice algorithm, also popular in the Middle Ages, is one that many of my students like.

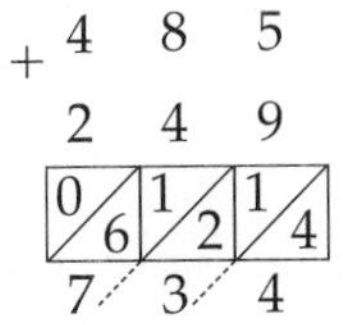

With this algorithm, you can begin at the left or the right. First, you find the sum of the digits in each place and put each of the sums in the box below, as shown. To determine the answer, you extend the diagonal line segment inside each box and then add the numbers that are in the same "chute," as shown above.

Copyright © Houghton Mifflin Company. All rights reserved.

EXPLORATION 3.4 Addition and Number Sense

You will be asked to grapple with some different kinds of problems for each of the operations. The purpose here is not to develop better computation or to understand why algorithms work. Rather the purpose is to develop number and operation sense that can be seen as common sense with numbers and operations.

In each problem, be prepared to explain your justification for your answers.

1. I'm thinking of 4 numbers whose sum is greater than 100.
   Tell whether each of the following must be true, might be true, or can't be true:
   a. All four numbers are greater than 20.
   b. If two of the numbers are less than 25, the other two must be greater than 25.
   c. All four numbers are two-digit numbers.

2. Determine the value of a, b, and c that will make this addition problem true:

$$\begin{array}{r} abc \\ +cba \\ \hline 625 \end{array}$$

3. Determine which of these answers are reasonable. You have about 5 seconds to make your determination. That is, make your determination, using number sense, without doing any pencil-and-paper work or trying to add the numbers mentally

$$\begin{array}{r} 40 \\ 38 \\ 42 \\ +\ 45 \\ 39 \\ 48 \\ 42 \\ \hline 340 \end{array} \qquad \begin{array}{r} 65{,}482{,}588 \\ +\ 47{,}546{,}716 \\ \hline 103{,}546{,}224 \end{array} \qquad \begin{array}{r} 999 \\ +\ 9{,}999 \\ 899 \\ \hline 10{,}627 \end{array}$$

Copyright © Houghton Mifflin Company. All rights reserved.

SECTION **3.2** **Exploring Subtraction**

EXPLORATION 3.5   **Mental Subtraction**

Most people find it more difficult to determine exact answers to subtraction problems in their heads. However, if you think about the various models for subtraction that we have discussed, there are many possibilities: take-away, comparison, adding up, and the fact that the difference between two numbers tells us how far apart they are on the number line, to name but a few.

1. Do in your head the six computations shown below. Briefly note the strategies that you used, other than simply doing the standard algorithm in your head. (Once again, this strategy is not bad; it's just one that everybody already has.) Try to give names to your strategies.

2. Share your strategies in your small group. Note any strategies that you heard that you did not use but would like to use.

3. In your group, select two or three strategies to describe to the class. Make up a name for each strategy.

4. After hearing the class presentations, write down the strategies that you like best.

Difference	What you did	Name
a. $\quad 65$ $-\ 28$		
b. $\quad 71$ $-\ 39$		
c. $\quad 80$ $-\ 36$		
d. $\quad 324$ $-\ 275$		
e. $\quad 152$ $-\ \ 87$		
f. $\quad 1000$ $-\ \ 378$		

Copyright © Houghton Mifflin Company. All rights reserved.

EXPLORATION 3.6 Subtraction: Children's Algorithms and Alternative Algorithms

PART 1: Children's algorithms

Below are descriptions of several solution paths for the problem $92 - 38$—paths that children commonly invent when they are not shown how to subtract, but, rather, are provided with rich questions to develop their understanding of place value and then are asked to figure out these problems on their own. In each case,

a. Make up and solve additional problems until you understand how the algorithm works.

b. Determine whether it will work for larger numbers: for $837 - 375$, for $904 - 268$, for $800 - 358$.

c. Write your first draft of why the algorithm works.

1. Subtract down.
   The child says: $92 - 30 = 62; 62 - 8 = 54$

2. Subtract the tens, and then do the ones separately.
   $90 - 30 = 60; 60 - 8 = 52; 52 + 2 = 54$

3. Subtract in each place; it's ok if you go negative.

   $$\begin{array}{r} 92 \\ -\ 38 \\ \hline 60 \\ -\ 6 \\ \hline 54 \end{array}$$

   $90 - 30 = 60$
   $2 - 8 = -6$
   $60 - 6 = 54$

4. Compensate: Add 2 to each number.
   $92 - 38$ becomes $94 - 40 = 54$

PART 2: Alternative algorithms

Below are several algorithms that have enjoyed popularity in different countries at different times. In each case,

a. Make up and solve more problems until you understand how the algorithm works.

b. Determine whether it will work for larger numbers: for $385 + 562$, for $476 + 508$, for $3245 + 683$.

c. Write your first draft of why the algorithm works.

d. Compare the algorithm to the standard addition algorithm in the text with respect to the five criteria Hyman Bass described, which are given on page 162 of the text.

The Indian Algorithm

This is one of the earliest algorithms for subtraction and was popular in India almost a thousand years ago. It is called the *reverse method* and has similarities to the standard algorithm used in the United States in that it involves a "borrowing" step and a "pay-back" step. We begin at the left and subtract the digits in the farthest left place. We proceed place by place. As long as no regrouping is required, we just sail along, as shown in Step 1. In this example, when we get to the ones place, we encounter the problem of not being able to subtract 4 from 2. As in the algorithm commonly used, we place a 1 above the 2 that is in the ones place, giving that place a value of 12. Now we do the subtraction $(12 - 4)$ and put the difference of 8 below. Then comes the payback: We cross

Copyright © Houghton Mifflin Company. All rights reserved.

EXPLORATION 3.21　Division and Number Sense

1.　Find the divisor.

$$\frac{65R19}{?\overline{)644}}$$

2.　Find the quotient

$$\frac{85R17}{56\overline{)?}}$$

3.　Is this answer reasonable? Why or why not? You have about 5 seconds to make your determination. That is, make your determination, using number sense, without doing any pencil-and-paper work or trying to multiply the numbers mentally.

$$\frac{704}{55\overline{)40000}}$$

4.　One way to estimate division problems is to round both numbers up or both numbers down. What if we were estimating 225/9? Which would be the better estimate: 250/10 or 230/10? Why?

Copyright © Houghton Mifflin Company. All rights reserved.

EXPLORATION 3.22   Developing Operation Sense

PART 1: Determining the appropriate operation

In each of the problems below, select the correct operations by a means other than random trial and error. Briefly explain the reasoning behind your guesses.

1. $2 \bigcirc 4 \bigcirc 3 = 11$
2. $43 \bigcirc 24 \bigcirc 68 = 1100$
3. $684 \bigcirc 418 \bigcirc 942 \bigcirc 246 = 962$
4. $(624 \bigcirc 319) \bigcirc (722 \bigcirc 699) = 41$

PART 2: Algebra and operation sense

In each of the cases below, the actual numbers have been replaced by variables. Write the number sentence that would correctly determine the answer.

1. Germaine bought A CDs for B dollars each. He sold each CD for C dollars. How much profit did he make?
2. Gary bought A boxes of golf balls each containing B balls. If he sold each box for C dollars, how much money did he make?
3. Mandy wants to buy a car for A dollars. She has currently saved B dollars. If she has C months to save up for the car, on average she needs to save D dollars each month.
4. Tasha is going on a trip of A miles. If her car gets about B miles per gallon, she will need C gallons of gas. If she figures that the gas will cost about D dollars per gallon, the cost of the gas for the trip will be E dollars.

PART 3: Which answer is reasonable?

Determine which of these answers are reasonable. You have about 5 seconds to make your determination. That is, make your determination, using number sense, without doing any pencil-and-paper work or trying to subtract the numbers mentally.

1. There were 35 students in a class and each student paid $28 for a field trip. The total cost was about $600.
2. There are 648 students in a school that has made a deal to buy each student a computer for $415 per student. The total cost of the program will be about $18,000.
3. The Boosters Club has to pay for the banquet. The meal cost $2143, the trophies cost $436, the senior gifts cost $562, and the raffle prizes cost $82. The club had $8265 money in the bank. After the banquet, the balance is about $5000.
4. A craftsperson has made 462 wooden spoons for the holiday season, and he will distribute these to 12 dealers. Each dealer will get about 25 spoons.

Copyright © Houghton Mifflin Company. All rights reserved.

EXPLORATION 3.23 Operation Sense in Games

Greatest amount

Materials

- One die with the following numbers: 1, 2, 3, 4, 5, 6.

Directions for playing the game

- Roll the die four times and record the numbers.
- The object is to make the value of the given expression as great as possible.

1. $\square \times \square + \square - \square$

   a. Before the first game, describe a master strategy that you think will work for all cases.
   b. Play several games.
   c. Write down your master strategy if it has changed.

2. $\square \times \square \div \square + \square$

   a. Do you think the same master strategy you developed in Step 1 will apply, or do you need to modify it?
   b. Play several rounds, recording your numbers and your reasoning.
   c. Write down your master strategy if it has changed.

3. $(\square \times \square)$

   $(\square - \square)$

   a. Before rolling the die, describe a master strategy for this game.
   b. Play several rounds, recording your numbers and your reasoning.
   c. Write down your master strategy if it has changed.

4. Play any of the games above. This time the goal is to make the value of the given expression as small as possible.

Target

This game is adapted from the *Fifth-Grade Book* in the NCTM Addenda Series (p. 48).

Materials

- Three dice:
  Die 1:   1, 2, 3, 4, 5, 6
  Die 2:   1, 2, 3, 4, 5, 6
  Die 3:   7, 8, 9, 10, 11, 12

Directions for playing the game

- Pick a target number.
- Roll the three dice.
- Use the three numbers you roll and any combination of operations to get as close as possible to the target number you have chosen.

1. For each target number and three numbers, write your answer and your reasoning.

Copyright © Houghton Mifflin Company. All rights reserved.

EXPLORATION 3.24 How Many Stars?

Without counting each star, determine how many stars are on this page. Describe and justify your process.

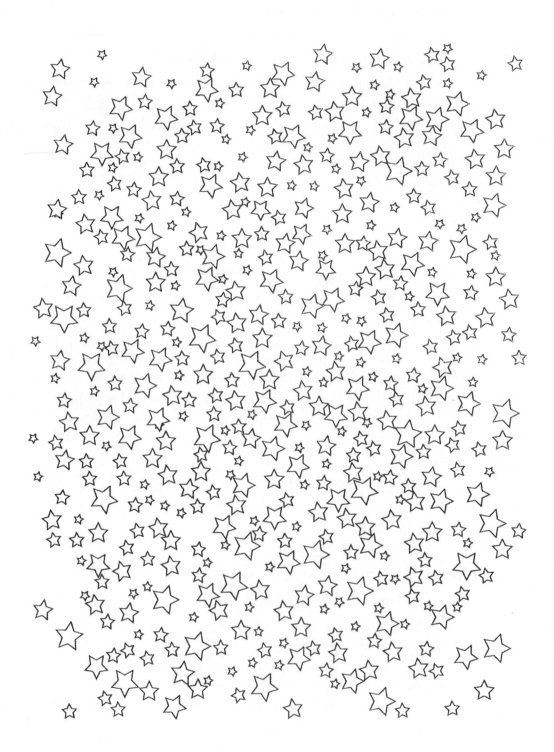

Copyright © Houghton Mifflin Company. All rights reserved.

4 Number Theory

*M*any people do not realize that mathematical concepts and ideas often have visual representations. For example, in Chapter 3 you investigated different visual representations of multiplication. In this chapter's explorations, you will be able to visualize some of the concepts that we will study in the text. These explorations can be done with elementary school children, yet they offer interesting challenges to the adult student too.

In each of the following explorations, you will not only lay the foundation for learning some important number theory ideas but will also have the opportunity to develop more mathematical tools—making and testing predictions. Students often tell me that when they think of mathematics, they think of numbers and computations. However, making and testing predictions has been crucial not only to the development of mathematics but also to the development of civilization. Young children are constantly making and testing predictions; I have cleaned up many messes my children made as a result of erroneous predictions! People make and test predictions regularly in their personal and professional lives. Yet somehow, this important human activity seems to be absent from much of school mathematics.

SECTION 4.1 Exploring Divisibility and Related Concepts

As you discovered in Chapter 3, being able to decompose and recompose a number in different ways is crucial to understanding computation algorithms for the four operations and to doing mental arithmetic and estimation confidently. Depending on the situation, we might decompose 48 as $40 + 8$, as $50 - 2$, as $8 \cdot 6$, or as $2 \cdot 2 \cdot 2 \cdot 2 \cdot 3$. In the following exploration, as you play the Taxman game more times, you will discover that one kind of decomposition of a number will enable you to increase your score, and you will see new relationships among numbers.

Copyright © Houghton Mifflin Company. All rights reserved.

EXPLORATION 4.1 Taxman

I don't know who invented this game, but I have seen variations in many different places. The version I present to you is one in which a team of two (or more) players competes against the "taxman."

Materials

- Array of consecutive numbers (see the game sheets on page 81).

Rules

1. The team crosses out a number.
2. The taxman crosses out all the proper factors of that number.
3. The team crosses out another number. A number can be crossed out only if at least one of its proper factors has not yet been crossed out. For example, you can't cross out 14 if 1, 2, and 7 have all been crossed out already. Play continues until none of the remaining numbers has a proper factor that has not been crossed out.
4. At the end of the game, the taxman gets all remaining numbers!
5. To determine your score, add up all the numbers your team crossed out.

1. Take out the Taxman Game Sheet on page 81 and play several games. As you are playing, write down observations about strategies. For example, is there a best first move? Is there a worst first move? Is it better to go for big numbers earlier or later? Listen to the class discussion about strategies, and write down anything you learned from that discussion.

2. Take out the Taxman Game Sheet for Step 2 on page 81 and play several games. As you are playing, write down observations about strategies. For example, is there a best first move? Is there a worst first move? Is it better to go for big numbers earlier or later? Listen to the class discussion about strategies, and write down anything you learned from that discussion.

3. Answer the following questions.

    a. If we were to play a game with a game sheet from 1 to 40, what would you recommend for the first move? Justify your choice.
    b. What would be the worst first move? Why?
    c. Let's say a team played the game with numbers 1–30 and their score was 234. To find the Taxman's score, they could add up all of the taxman's numbers, but there is an easier way to deduce the taxman's score. Can you find it?

4. As you have seen, some numbers have many factors and some have very few. The ancient Greeks were fascinated by relationships among numbers, and they came up with three terms to classify numbers with respect to their factors. If you add the sum of the proper factors of a number, that sum might be greater than, equal to, or less than the number itself. In the former case, a number is said to be abundant, in the second case perfect, and in the third case deficient.

    a. Classify the first 30 numbers in this manner.
    b. Write down any observations from your work.
    c. Predict what group the following numbers would be in, and briefly describe your prediction:

    > 47, 48, 49, 50

    d. Predict a number above 50 that you think is likely to be deficient, and explain your choice.
    e. Predict a number above 50 that you think is likely to be abundant, and explain your choice.
    f. How does this work connect to the game or help you with strategies for it?

Copyright © Houghton Mifflin Company. All rights reserved.

Taxman Game Sheet for EXPLORATION 4.1

1	2	3	4	5	6	7	8	9	10	Your score ____
11	12	13	14	15	16	17	18	19	20	Taxman ____

1	2	3	4	5	6	7	8	9	10	Your score ____
11	12	13	14	15	16	17	18	19	20	Taxman ____

1	2	3	4	5	6	7	8	9	10	Your score ____
11	12	13	14	15	16	17	18	19	20	Taxman ____

Taxman Game Sheet for EXPLORATION 4.1, Step 2

1	2	3	4	5	6	7	8	9	10	
11	12	13	14	15	16	17	18	19	20	
21	22	23	24	25	26	27	28	29	30	Score ____

1	2	3	4	5	6	7	8	9	10	
11	12	13	14	15	16	17	18	19	20	
21	22	23	24	25	26	27	28	29	30	Score ____

1	2	3	4	5	6	7	8	9	10	
11	12	13	14	15	16	17	18	19	20	
21	22	23	24	25	26	27	28	29	30	Score ____

Copyright © Houghton Mifflin Company. All rights reserved.

SECTION **4.2** Exploring Prime and Composite Numbers

EXPLORATION **4.2** Factors

An exploration many elementary teachers use to help students develop number sense is to challenge them to make as many different rectangles as they can using a certain number of unit squares. For example, how many different rectangles can be made with six squares?

We can make two different rectangles with six unit squares. If we represent them by their length and width, we have a 6 × 1 rectangle and a 3 × 2 rectangle. Using mathematical terminology, we can also say that 6 has these four factors: 1, 2, 3, 6.

When elementary school children investigate all the different rectangles they can make for each number, the exploration not only reinforces their multiplication facts but also addresses other important areas: problem-solving, communication, reasoning, and making connections. Because you are able to think at a more abstract level, we will modify the instructions given to the children.

1. Take out the Factors Table on page 85. In the table, list the factors of each of the first 25 natural numbers.

2. Describe any observations, hypotheses, and questions you have as a result of filling out and looking at the table.

3. One of the themes of this book is the power of different representations.

   a. Take out the Number of Factors Table on page 87. Use the data in your Factors Table from Step 1 to complete the table.

   b. Note any observations, hypotheses, and questions that you have as a result of filling out and looking at this table.

   c. In one sense, the numbers in each column of the table constitute a "family." Can you give a name to any of the families? If so, write down this name (it does not have to be a "mathematical" name) and describe the characteristics that are common to all members of the family.

   d. Suppose we were to continue to look at the factors of numbers beyond 25. Predict (and briefly describe your reasoning) the next number that will have 2 factors, the next number that will have 3 factors, and the next number that will have 4 factors.

   e. Now extend the table until you have the next number that has 2 factors, 3 factors, and 4 factors. Describe any insights or discoveries from the extension.

   f. Can you predict the first number that will have 7 factors? What will it look like? Justify your reasoning.

   g. There are other possible families of numbers, based on the number of factors. For example, if we combine the 3-factor families, the 5-factor families, and the 7-factor families into a large family called "odd number of factors," how might you describe the characteristics that are common to all members of *this* family?

Copyright © Houghton Mifflin Company. All rights reserved.

Factors Table for EXPLORATION 4.2, Step 1

Number	Factors							
1	1							
2	1	2						
3	1	3						
4								
5								
6								
7								
8								
9								
10								
11								
12								
13								
14								
15								
16								
17								
18								
19								
20								
21								
22								
23								
24								
25								

Copyright © Houghton Mifflin Company. All rights reserved.

Number of Factors Table for EXPLORATION 4.2, Step 3

1 factor	2 factors	3 factors	4 factors	5 factors	6 factors	7 factors	8 factors	9 factors
1	2	4						
	3							
	5							

Copyright © Houghton Mifflin Company. All rights reserved.

6. Many children have had fun with this one. The directions are to write your name on graph paper and see the patterns that occur with different lengths. You are finished when the last letter of your name appears on the last column, thus making a rectangle. The case for a 3-letter name is shown below for three different lengths.

4 columns				5 columns					7 columns						
P	A	T	P	P	A	T	P	A	P	A	T	P	A	T	P
A	T	P	A	T	P	A	T	P	A	T	P	A	T	P	A
T	P	A	T	A	T	P	A	T	T	P	A	T	P	A	T

a. Using a sheet of "Other Base Graph Paper" at the end of this book, try your name with various lengths. Describe any patterns you see, your observations, and any questions.

   Two questions that often come from the initial exploration are described below. Exploring them involves some wonderful opportunities to understand various aspects of number theory, as well as to develop your competence with the five process standards stated by NCTM. Read the two questions and then follow the directions immediately below the questions.

b. *Question 1* Background: You will have noticed that in some cases, there is a diagonal pattern going from bottom left to top right (as in the 4- and 7-column example) above, sometimes there is a diagonal pattern going the other way (as in the 5-column example), and sometimes there is no real diagonal pattern.

   The question: Can you predict when a rectangle will have the first diagonal pattern, the second diagonal pattern, or no diagonal pattern, given the length of letters in the "name" and the length of the column?

c. *Question 2* Background: We can also look at the dimensions of the rectangle formed by one complete cycle. In the case of a 3-letter name, the rectangle for a 4-column pattern is 4×3, the rectangle for a 5-column pattern is 5×3, and the rectangle for a 7-column pattern is 7×3.

   The question: Can you predict the dimensions of the rectangle, given the length of letters in the "name" and the length of the column? Note that there are many possibilities explaining the makeup of the finished rectangle—odd numbers, even numbers, prime numbers, numbers with common factors, and many more!

Copyright © Houghton Mifflin Company. All rights reserved.

EXPLORATION 5.4 Understanding Integer Division

Repeated-Subtraction and Partitioning Models

Consider the problem $8 \div 2$. We learned in Chapter 3 that we can represent this problem using the partitioning model of division, and we can represent this problem using the repeated-subtraction model of division. For review, you may want to model the problem $8 \div 2$ with the partitioning model and with the repeated-subtraction model.

1. **a.** Model the problem $^-8 \div 2$ with the partitioning model.
   b. Model the same problem with the repeated-subtraction model.
   c. Describe and justify your process so that a reader who is not familiar with integer division could understand both how you got the answer and why your process works.

Missing-Factor Model

When we get to other cases of integer division (for example, $8 \div {}^-2$ and $^-8 \div {}^-2$), these two models become more and more complex, and at some point, they pass into the realm of "more trouble than it's worth." Therefore, let us abandon the partitioning and repeated-subtraction models for division and look to another model for division to justify the procedures for integer division: the missing-factor model. When we use this model, we find the solution to $a \div b$ by asking, "What number times b is equal to a?"

2. Use the missing factor model to determine the answers to the following three problems. As before, describe and justify your process so that a reader who is not familiar with integer division could understand both how you got the answer and why your process works.

$$^-8 \div 2$$

$$12 \div {}^-3$$

$$^-20 \div {}^-4$$

3. We have now examined all four cases that can occur when dividing integers. On the basis of your work above, can you describe a general rule for dividing integers?

Copyright © Houghton Mifflin Company. All rights reserved.

SECTION **5.2**   Exploring Fractions and Rational Numbers

The fraction explorations here have been designed to give you an opportunity to work with the fundamental concepts related to fractions. Just as we discussed different decompositions of whole numbers in Chapters 3 and 4, a key to understanding fractions is to decompose them. That is, the notion of parts and wholes connects to our compositions and decompositions with whole numbers. We shall investigate various decompositions, which in turn will deepen your understanding of the different fraction contexts and the relationship between the numerator and the denominator.

EXPLORATION 5.5   Egyptian Fractions and Sub Sandwiches

When we ask children to explore, their natural solution paths often mirror strategies developed over the centuries in different parts of the world.

The problem that prompted this exploration is this: How can 8 kids share 7 sub sandwiches? One solution path, shown at the right, is to figure out the biggest piece that all can share and then move to smaller and smaller pieces. For example, first give everyone 1/2 a sandwich. Then give everyone 1/4 of a sandwich. Now you have one sandwich left, so give everyone 1/8 of a sandwich. Everyone gets $1/2 + 1/4 + 1/8$. This process is very similar to how the Egyptians worked with fractions. It turns out that any fraction can be decomposed into some combination of different unit fractions.

This exploration asks you to determine each person's share in the following division problems. A valid answer consists of a sum of unit fractions, and all of the fractions need to be different; for example, $7/10 = 1/2 + 1/10 + 1/10$ is not valid, but $7/10 = 1/2 + 1/5$ is valid. Look for patterns and strategies that will enable you to be more effective. In most of the cases, the solution will involve the sum of 2 fractions. In the other cases, it will involve the sum of 3 fractions.

1. How would 5 children share 2 sub sandwiches?
2. How would 5 children share 3 sub sandwiches?
3. How would 5 children share 4 sub sandwiches?
4. How would 6 children share 4 sub sandwiches?
5. How would 6 children share 5 sub sandwiches?
6. How would 7 children share 2 sub sandwiches?
7. How would 7 children share 3 sub sandwiches?
8. How would 7 children share 4 sub sandwiches?
9. How would 7 children share 5 sub sandwiches?
10. How would 7 children share 6 sub sandwiches?
11. How would 8 children share 5 sub sandwiches?
12. How would 8 children share 6 sub sandwiches?
13. Make up and solve your own problems.

Copyright © Houghton Mifflin Company. All rights reserved.

14. Describe any shortcuts or faster strategies that you developed while doing these problems.

15. Describe any growth in your understanding of fractions (such as your understanding of equivalent fractions, factors, or LCM) that resulted from doing these problems.

16. Can you generalize your approach to an algorithm; that is, can you describe a procedure that is general enough that it would work for any combination of children and sandwiches?

Copyright © Houghton Mifflin Company. All rights reserved.

EXPLORATION 5.6   Making Manipulatives

Using manipulatives in elementary school mathematics was one of the bandwagons of the 1980s. In fact, "hands-on" became one of those infamous buzzwords. However, hands-on alone is not enough. One modification of this phrase is "hands-on and minds-on." In other words, although manipulatives can help to ground one's understanding of new ideas, what the hands do needs to be connected to important ideas in the mathematical concept.

One of the reasons for the limited success of much work with manipulatives on fractions is that even when students have worked with manipulatives, they have often worked with premade manipulatives. Although there is a place for premade manipulatives in the curriculum, many mathematics educators believe that at some point early in their development, students will benefit tremendously by making their own manipulatives. Such an activity not only allows the students to grapple with concepts at a deeper level but also encourages more creativity on their part. The following exploration has been one of my students' favorites.

Making the Manipulatives

Some of you will make fraction manipulatives from circles and some from squares. Regardless of which manipulatives you make, read the questions in Steps 1–3 below and keep them in mind as you make your manipulatives.

Circles: Using construction paper that has been cut into circles of the same size, make a set of manipulatives. The only restriction is that you cannot use a protractor. For example, you cannot make 1/4 circles by measuring 90-degree angles. In addition to physical tools, you will also need to use a combination of problem-solving strategies, including reasoning and guess–check–revise.

Squares: Using construction paper that has been cut into squares of the same size, make a set of manipulatives. The only restriction is that you cannot use a ruler to measure the divisions. For example, if your square is 6 inches on a side, you cannot use the ruler to make a mark every 2 inches. In addition to physical tools, you will also need to use a combination of problem-solving strategies, including reasoning and guess–test–revise.

After you have made your manipulatives, respond to the following questions.

1. **a.** Describe how you made thirds as though you were talking to someone who missed class. Your description needs to have enough specificity so that the person could repeat what you did and see why you did it that way.
    b. Describe how you made fifths as though you were talking to someone who missed class.

2. Describe at least two learnings that resulted from making your set of manipulatives.

3. Describe any questions that you have at this point, either questions about how to make a particular fraction (such as ninths) or other questions that arose when you were making your sets.

4. Alicia said that making 1/5 was hard. Brandon said that it was easy because 1/5 is "halfway between 1/4 and 1/6." Jamie said that Brandon's method can be used for lots of cases; for example, 1/7 will be halfway between 1/6 and 1/8, and 1/6 will be halfway between 1/4 and 1/8.

    a. What is your initial reaction to Brandon's statement? Do you feel that 1/5 is halfway between 1/4 and 1/6 or not? Why?
    b. Discuss this question in your group. If you changed your mind, explain what changed your mind and justify your present position. If you didn't change your mind, but you feel you can better justify your response, write your revised justification.

5. Leah called some fractions "prime fractions." What do you think she meant?

Copyright © Houghton Mifflin Company. All rights reserved.

EXPLORATION 5.7   Wholes and Units: Not Always the Same

Thinking of fractions only in terms of parts and wholes is simplistic and causes problems when we get beyond simple, routine problems. The following two problems will help you to grapple with some very important fraction ideas.

1. *Absent Students*[2]   Let's say you are teaching. It is winter, and it seems that a larger fraction of students than normal are sick. You are eating lunch in the teachers' lounge, and another teacher says, "In my class today, 2/5 of the girls were absent but only 1/5 of the boys were absent." What fraction of her class was absent today?

   a. Work on the problem above and show your work.
   b. What if the number of boys and girls is equal? Does this change your answer in a. or not? Explain your response.
   c. Meet with your partner(s) and discuss the problem. If you have the same answer, did you arrive at the answer in the same way? If you have different answers, listen to one another's reasoning until you can agree on one answer. Justify your answer.
   d. What is a realistic range of possible answers?
   e. What is the theoretical range of possible answers?

2. *How Much Is Her Share?*[3]   This problem is adapted from a real-life problem. Josephine is a graduate student at Urban State College. Because her financial resources are limited, she has moved into a house with four other people. The house is heated with electricity, and the electric bill comes every two months. Josephine moved in on February 1. When the bill for January–February comes, what fraction of the bill should each person pay?

   a. Work on this problem alone and show your work.
   b. What assumptions did you make in order to solve this problem?
   c. Meet with your partner(s) and discuss the problem. If you have the same answer, did you arrive at the answer in the same way? If you have different answers, listen to one another's reasoning until you can agree on one answer. Justify your answer.

Copyright © Houghton Mifflin Company. All rights reserved.

[2]This problem has been adapted from one developed by Deborah Schifter at Education Development Center.
[3]This problem has been adapted from a problem in *A Course Guide to Math 010L*, by Ron Narode, Deborah Schifter, and Jack Lochhead, 1985.

EXPLORATION 5.8   Fractions with Different Manipulatives

There are several parts to this exploration, which will help you explore fundamental fraction ideas and operations using three different manipulatives: Pattern Blocks, Cuisenaire rods, and Geoboards.

PART 1: The meaning(s) of fractions

The following questions have been designed to make you think about what the numerator and denominator mean and what the term *fraction* means.

1. Whole numbers have a very concrete, tangible meaning; for example, 3 means this many (x x x) of something. However, 1/2 and other fractions do not have that same kind of concrete meaning. Do the following activities, and then describe what 1/2 means that works for all of the cases below, along with any additional examples you can think of.

    Pattern Blocks: If ⬡ = 1, then ⬭ = $\frac{1}{2}$. However, if ◇ = 1, then △ = $\frac{1}{2}$.

    Cuisenaire rods: If ▭ = 1, then ▢ = $\frac{1}{2}$. However, if ▭▭▭▭▭▭ = 1,

    then ▭▭▭ = $\frac{1}{2}$.

    Geoboards: In both geoboards, if the big region = 1, then the smaller region = $\frac{1}{2}$.

    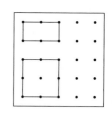

2. **a.**   Make 5/6 with each manipulative, and sketch it on your paper.
    b.   Now explain what 5/6 means. Can you use your explanation for 1/2, just substituting 6 for 2 and 5 for 1? If so, why? If not, write an explanation of the term *fraction* that will work in both cases.

    For Problems 3–5, answer each question, sketch your solution, and briefly explain your solution path—that is, how you got your answer.

3. If ⬡⬡ = 1, sketch $\frac{2}{3}$.

    If ▭▭▭▭▭▭ = 1, sketch $\frac{2}{3}$.

    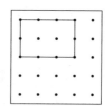

    If the rectangle at the right = 1, draw a picture of $\frac{2}{3}$.

    Think what is similar in all these cases and then write directions for solving "if [blank] = 1, then sketch $\frac{a}{b}$" that will work for all three manipulatives.

Copyright © Houghton Mifflin Company. All rights reserved.

4. If $\triangle = \dfrac{3}{4}$, sketch 1.

If [rectangle divided into 8 squares] $= \dfrac{3}{4}$,

sketch 1.

If the rectangle $= \dfrac{3}{4}$, draw a picture of 1.

Now write directions for solving "if [blank] $= \dfrac{a}{b}$, then sketch 1" that will work for all three manipulatives.

5. If [two hexagons] $= 1\dfrac{1}{3}$, sketch 1.

If [two rectangles of squares] $= 1\dfrac{1}{3}$,

sketch 1.

If the rectangle $= 1\dfrac{1}{3}$, draw a picture of 1.

Now write directions for solving "if [blank] $= a\dfrac{b}{c}$, then sketch 1" that will work for all three manipulatives.

6. Make up a problem of your own and solve it. Challenge yourself!

7. One goal of Part 1 is for you to realize that we have to be careful about the word *whole*. In one case above, the whole was equal to the unit (i.e., 1), but in another case, the whole was 3/4, and in the last case, the whole was 1 1/3. Similarly, to make sense of 5/4, you can't say, "The whole has been divided into 4 pieces and we have 5 of them."

    a. Describe what *whole* and *unit* mean with respect to working with fractions, as though speaking to a child who is struggling with fractions.

    b. Go back and review your definition of *fraction* from Step 2. Does it still work? If so, why do you still like it? If not, revise it so that it works. Can you define *fraction* without using the word *whole*?

PART 2: Ordering fractions

1. Determine which fraction in each pair is greater, first with Pattern Blocks, then with Cuisenaire rods, and then with Geoboards. In each case, show your work, including sketches, and then briefly explain your solution.

    a. 5/6 or 2/3?   **b.** 3/4 or 5/6?   **c.** 4/5 or 2/3?

2. Describe the problem that was hardest for you. Explain what was hard about it and describe what unlocked the problem for you.

3. Were any of the manipulatives easier to use than others? If so, describe what manipulative was easiest for you and why.

4. As earlier, we now focus on generalizing. Imagine being given two fractions a/b and c/d with reasonable denominators. Write directions for deciding which is greater that will work for all three manipulatives.

Copyright © Houghton Mifflin Company. All rights reserved.

PART 3: Equivalent fractions

1. Use each manipulative to demonstrate that 2/3 = 8/12. Sketch your work and explain your reasoning.

2. Describe any differences in how the equivalence of 2/3 and 8/12 was demonstrated or justified.

3. Now justify the algorithm for making equivalent fractions. This means explain why it works, not how it works.

$$\frac{2}{3} = \frac{2 \cdot 4}{3 \cdot 4} = \frac{8}{12}$$

PART 4: Adding fractions

1. For the moment, I would like you to suspend your knowledge that you have to get a common denominator to add fractions. Imagine that you don't have any algorithms and thus have to find the answers using the manipulatives and your knowledge of what *fraction* means. Using your manipulatives, add the following fractions and briefly explain your reasoning.

   a. $\frac{1}{2} + \frac{1}{3}$ **b.** $\frac{1}{4} + \frac{2}{3}$ **c.** $\frac{3}{4} + \frac{1}{6}$ **d.** $1\frac{2}{3} + 2\frac{3}{4}$

2. On the basis of your work above, explain why we have to find a common denominator in order to add fractions. Some students find this alternative version of the question preferable: Why can't we just add the top numbers and add the bottom numbers? For example, why isn't $\frac{3}{4} + \frac{1}{6} = \frac{4}{10}$?

Looking Back on Exploration 5.8

1. Look back on this exploration, think about fractions and the meaning of various fraction terms (numerator, denominator, whole, and unit), and think about fractions with different models. Describe any growth in your understanding of fractions.

2. What questions do you have about fractions?

Copyright © Houghton Mifflin Company. All rights reserved.

EXPLORATION 5.9 Partitioning Wholes

PART 1: Understanding area models

1. Divide each rectangle into the number of pieces specified.

    a. 3 equal pieces

    b. 4 equal pieces

    c. 8 equal pieces

2. Divide each Geoboard into 2 equal pieces. Your solution cannot involve the simple solution of a rubber band down the middle. Be creative!

3. Divide each triangle into 3 equal pieces.

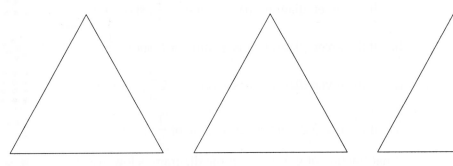

Copyright © Houghton Mifflin Company. All rights reserved.

4. **a.**   If the rectangle has a value of $\frac{1}{3}$, show 1.

   b.   If the rectangle has a value of $\frac{3}{4}$, show 1.

   c.   If the rectangle has a value of $1\frac{1}{3}$, show 1.

5. The following questions are related to Pattern Blocks.

   a.   If the hexagon has a value of $\frac{2}{3}$, show 1.

   b.   If the trapezoid has a value of $\frac{3}{4}$, show 1.

   c.   If two hexagons have a value of $1\frac{1}{4}$, show 1.

   d.   If the trapezoid has a value of $\frac{3}{4}$, what is the value of the rhombus?

6. If each of the hexagons has a value of 1, does each small piece in each hexagon represent $\frac{1}{6}$? Why or why not?

7. Is each of the arrays below a valid representation of $\frac{4}{3}$? Why or why not?

PART 2: Understanding discrete models

1. Draw a diagram that has the specified value. Justify your solutions.

   a.   If the given diagram has a value of $\frac{2}{3}$, show 1.

   b.   If the given diagram has a value of 1, show $\frac{3}{4}$.

   c.   If the given diagram has a value of $2\frac{2}{3}$, show 1.

   d.   If the given diagram has a value of $\frac{1}{3}$, show $\frac{1}{2}$.

2. What fraction of the circles in the diagram below are black? Your answer needs to have a denominator other than 12. Justify your answer.

Copyright © Houghton Mifflin Company. All rights reserved.

PART 3: Understanding number line models

1. **a.**   On the number line below, mark the approximate location of 1. Explain your reasoning.

$$\underset{\substack{\;\\0}}{\vdash}\underset{\substack{\;\\ \frac{1}{3}}}{\vert}\longrightarrow$$

    b.   On the number line below, mark the approximate location of 1. Explain your reasoning.

$$\underset{\substack{\;\\0}}{\vdash}\qquad\underset{\substack{\;\\ \frac{3}{4}}}{\vert}\longrightarrow$$

    c.   On the number line below, mark the approximate location of 1. Explain your reasoning.

$$\underset{\substack{\;\\0}}{\vdash}\qquad\qquad\underset{\substack{\;\\ 1\frac{2}{3}}}{\vert}\longrightarrow$$

    d.   On the number line below, approximate the value of *x*. Explain your reasoning.

$$\underset{\substack{\;\\0}}{\vdash}\qquad\underset{\substack{\;\\ x}}{\vert}\quad\underset{\substack{\;\\ 1}}{\vert}\longrightarrow$$

    e.   On the number line below, approximate the value of *x*. Explain your reasoning.

$$\underset{\substack{\;\\0}}{\vdash}\quad\underset{\substack{\;\\ x}}{\vert}\qquad\qquad\underset{\substack{\;\\ 6}}{\vert}\longrightarrow$$

Looking Back on Exploration 5.9

1.   Which models were easier for you to work with? Why?
2.   Which models were harder for you to work with? Why?
3.   Describe your most important learnings.
4.   Describe any problems that still puzzle you.

Copyright © Houghton Mifflin Company. All rights reserved.

EXPLORATION 5.10 Ordering Fractions[4]

In this exploration, you will be asked to use your knowledge of fractions and your reasoning to order fractions. We will ask you to do this *without finding the LCM or using a calculator to convert the fractions to decimals* because the goal here is not simply to get the right answer but also to apply basic fraction ideas. Your justifications need to go beyond drawing a picture of the two fractions and saying that one looks bigger. For example, 5/9 is greater than 4/9 because both fractions have the same denominator, and hence all of the pieces are the same size. Therefore, having 5 such pieces is more than having 4 such pieces.

Take out the tables on pages 111–112, for use in Steps 1 and 4 below.

1. Use the table on page 111. Predict the relative values of the fractions by inserting the symbol <, =, or > into the space between the fractions, using fraction ideas without finding the LCM or converting to decimals. Briefly justify your choice.

2. Compare your responses with your partner(s). In some cases, you may have the same answer but different justifications.

3. With your partner(s), develop a set of general rules that could be used by another person. There are many possible rules and many ways to state them. For example, if two fractions have the same denominator, the fraction with the larger numerator is greater. Justify at least two of your rules; that is, explain *why* they work.

4. Exchange rules with another group. Then use *the other group's rules* to order the fractions in the table on page 112.

5. Critique the other group's rules with respect first to validity and then to clarity. That is, if you feel that any of the rules are invalid, explain why. If you feel that any rules are not clear, circle the words or phrases that are ambiguous or unclear and explain why you find them so.

6. On the basis of the critique from the other group, make any necessary changes in your rules.

7. Order each set of fractions again without finding the LCM for the set or converting them to decimals. Explain and justify your thinking process.

a. $\dfrac{31}{80}$ $\dfrac{13}{17}$ $\dfrac{2}{3}$

b. $\dfrac{1}{3}$ $\dfrac{4}{7}$ $\dfrac{2}{5}$ $\dfrac{7}{8}$ $\dfrac{5}{16}$

c. $\dfrac{3}{10}$ $\dfrac{2}{3}$ $\dfrac{7}{12}$ $\dfrac{4}{5}$ $\dfrac{3}{7}$

d. $\dfrac{1}{8}$ $\dfrac{2}{5}$ $\dfrac{5}{8}$ $\dfrac{5}{6}$ $\dfrac{3}{49}$ $\dfrac{3}{56}$

[4]This exploration has been adapted from one developed by the Summermath for Teachers program at Mt. Holyoke College and the Mathematics for Tomorrow project at Education Development Center.

Copyright © Houghton Mifflin Company. All rights reserved.

Table for EXPLORATION 5.10, Step 1

	Fraction	>, =, or <	Fraction	Justification
a.	$\frac{3}{5}$		$\frac{3}{8}$	
b.	$\frac{5}{6}$		$\frac{7}{8}$	
c.	$\frac{3}{5}$		$\frac{5}{12}$	
d.	$\frac{1}{2}$		$\frac{17}{31}$	
e.	$\frac{3}{8}$		$\frac{2}{9}$	
f.	$\frac{2}{7}$		$\frac{3}{8}$	
g.	$\frac{1}{4}$		$\frac{2}{9}$	
h.	$\frac{9}{11}$		$\frac{7}{9}$	
i.	$\frac{3}{8}$		$\frac{4}{10}$	
j.	$\frac{3}{10}$		$\frac{9}{23}$	

Copyright © Houghton Mifflin Company. All rights reserved.

Table for EXPLORATION 5.10, Step 4

	Fraction	>, =, or <	Fraction	Justification
a.	$\dfrac{3}{4}$		$\dfrac{7}{12}$	
b.	$\dfrac{5}{8}$		$\dfrac{10}{13}$	
c.	$\dfrac{5}{12}$		$\dfrac{7}{13}$	
d.	$\dfrac{7}{10}$		$\dfrac{14}{19}$	
e.	$\dfrac{2}{7}$		$\dfrac{1}{3}$	
f.	$\dfrac{3}{8}$		$\dfrac{4}{7}$	

Copyright © Houghton Mifflin Company. All rights reserved.

<u>**EXPLORATION 5.17**</u> **Exploring Decimal Algorithms**

You know *how* to add, subtract, multiply, and divide decimals. This exploration is to develop your understanding of and ability to explain *why* those algorithms work.

1. Addition
    a. Below are two examples of common mistakes made by children. What might the children be thinking that would explain why they do it this way?

Problem	Wrong answer	Problem	Wrong answer
$34 + 4.2$	$\begin{array}{r} 3\,4 \\ +4.2 \\ \hline 7.6 \end{array}$	$20.45 + 7.6$	$\begin{array}{r} 2\,0\,.4\,5 \\ +7.6 \\ \hline 9.6\,4\,5 \end{array}$

    b. In your own words, explain why we need to line up the decimal points when adding decimals.
    c. Explain how understanding of place value helps to understand the algorithm.
    d. Explain how understanding of fractions helps to understand the algorithm.

2. Multiplication
    In Chapter 3, you learned that various algorithms for multiplication automatically put the digits of the partial products into the right place. So, too, for multiplication with decimals. We determine the placement of the decimal point by adding the number of decimal places in the multiplicand and the multiplier. Why does this work? That is, why does this algorithm result in all of the digits being in the right place? You can use either or both of the following examples if you wish.

$$\begin{array}{r} 3.6 \\ \times\ \ 2.4 \\ \hline 1\,4\,4 \\ 7\,2\ \ \\ \hline 8.6\,4 \end{array} \qquad \begin{array}{r} 4.3\,6 \\ \times\ \ \ 7.8 \\ \hline 3\,4\,8\,8 \\ 3\,0\,5\,2\ \ \\ \hline 3\,4.0\,0\,8 \end{array}$$

3. Division
    a. To divide decimals, you know to move the decimal points in the divisor and the dividend so that the divisor is a whole number. The placement of the decimal point in the quotient is directly above the decimal point in the dividend. Explain why this algorithm works.

$$7.4\overline{)247.9} \qquad \begin{array}{r} 33.5\ \ \\ 74\overline{)2479.0} \\ \underline{222}\ \ \ \ \ \ \\ 259\ \ \\ \underline{222}\ \ \\ 370 \\ \underline{370} \end{array}$$

    b. Some children can perform this algorithm when the divisor has fewer decimal places than the dividend, but then they flounder when the divisor has more decimal places than the dividend. Below are two such examples. Why can we just add zeros to the dividend?

$$6.2\overline{)245} \qquad \rightarrow \qquad 6.2\overline{)245.0} \qquad \rightarrow \qquad 62\overline{)2450.}$$

$$7.54\overline{)46.3} \qquad \rightarrow \qquad 7.54\overline{)46.30} \qquad \rightarrow \qquad 754\overline{)4630.}$$

Copyright © Houghton Mifflin Company. All rights reserved.

EXPLORATION 5.18 Patterns in Repeating Decimals

What we convert some fractions to decimals, sometimes the decimal terminates and sometimes it repeats. For example, $1/16 = .0625$, but $1/11 = .0909090909090909 \ldots$ Rather than writing the "dot dot dot," we place a bar over the portion that repeats, and we present $1/11$ as $0.\overline{09}$.

Note: You might choose to use a spreadsheet to assist in the computations.

PART 1: Predicting the next number in a sequence

1. **a.** The decimal equivalent of $1/11$ is $0.090909\ldots$. Predict the decimal equivalent of $2/11$ and explain your reasoning. Then calculate $2/11$. If your prediction was correct, great. If not, examine the answer so that you can predict the value of $3/11$.
    b. Predict the value of $3/11$ and explain your reasoning. Determine the value of $3/11$. If your prediction was correct, great. If not, determine the value of $4/11, 5/11$, and so on, until you do see a pattern.
    c. Describe the pattern you see in words.
    d. Predict the value of $17/11$. Explain your reasoing.

2. Determine the decimal equivalent of $1/15$.

    a. Repeat the process described in the first problem:
    Predict and determine the next amount until you can describe the pattern.
    Describe the pattern you see in words.
    b. Predict the value of $32/15$. Explain your reasoning.

3. Determine the decimal equivalent of $1/99$.

    a. Repeat the process described in the first problem:
    Predict and determine the next amount until you can describe the pattern.
    Describe the pattern you see in words.
    Check other fractions in the pattern to see how long the pattern holds.
    b. Predict the value of $47/99$. Explain your reasoning.
    c. Based on your work in the previous problem, predict the value of $1/999$. Explain the reasoning behind your prediction. Then determine the decimal equivalent of $1/999$ and repeat the process used in the previous problems.
    d. Predict the value of $74/999$.

4. Determine the decimal equivalent of $1/101$.

    a. Repeat the process described in the first problem:
    Predict and determine the next amount until you can describe the pattern.
    Describe the pattern you see in words.
    Check other fractions in the pattern to see how long the pattern holds.
    b. Predict the value of $52/101$. Explain your reasoning.
    c. Based on your work in the previous problem, predict the value of $1/1001$. Explain the reasoning behind your prediction. Then determine the decimal equivalent of $1/1001$ and repeat the process used in the previous problems.
    d. Predict the value of $64/1001$.

5. Determine the decimal equivalents of $1/7, 2/7, 3/7, 4/7, 5/7, 6/7$. Describe the pattern in this sequence of decimals.

6. Predict other families of fractions that might be repeating fractions (e.g., $1/11, 1/18$, etc.) Check them out. Describe what kinds of denominators you think might result in repeating fractions, such as odd numbers, and so forth.

Copyright © Houghton Mifflin Company. All rights reserved.

PART 2: When will a decimal terminate and when will it repeat?

Now we will investigate a question which fascinated mathematicians several hundred years ago: Can we predict whether the decimal form of a fraction will terminate or repeat?

1. Write down your present thoughts as hypotheses. For example, it will terminate if _____ ; it will repeat if _____ .

2. Determine a plan. For example, you might proceed systematically: $\frac{1}{2}, \frac{1}{3}, \frac{1}{4}$, etc. You might proceed by groups of fractions with common characteristics, such as the denominator is an odd number or prime number, and so on.

3. Collect your data. You might choose to use a spreadsheet to be able to check your hypotheses more quickly. To maximize the development of your mathematical power, use a process like the one shown below. That is, select a fraction, predict whether you think the decimal form will repeat or terminate, and your reasoning. If your prediction was wrong, describe what you learned from analyzing your prediction.

Number	Prediction	Analysis/insight
1/13	Repeat because odd number	

4. Present and justify your findings.

Copyright © Houghton Mifflin Company. All rights reserved.

EXPLORATION 5.19 The Right Bucket: A Decimal Game

Materials

- Game sheet, page 125.
  After playing one or more games, answer the following questions.

1. What did you learn about estimating and mental math with decimals from this game? If you learned specific strategies, describe them.

2. Is it possible to choose the pairs of decimals so that you can get 3 points each time? If you think so, explain why. If you think not, explain why not.

The value of this game comes from refining your ability to estimate. This game comes from *Ideas from the Arithmetic Teacher: Grades 6–8*.[10]

Directions for playing the game

1. Select two decimals from the list given on the game sheet and cross them off the list.

2. Estimate their product and explain your reasoning. An example is shown in the table below.

3. Multiply the two decimals.

4. Determine your score by looking at the bucket chart below. For example, products between 10 and 100 earn a score of 2 points.

5. Select two more decimals, find their product, and determine your score. Continue until you have used all the decimals.

6. The goal is to score as many points as you can.

Bucket Chart

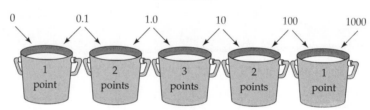

Note: This game can be played competitively (alternating turns) or cooperatively (working together to determine which two decimals make the best pick).

Turn	Decimals	Estimate	Reasoning	Actual product	Points
1	41.2	4.12	0.083 is almost 1/10, and I knew 1/10 of 41.2 is 4.12. Thus, I felt that 0.083 of 41.2 would be well over 1.	3.4196	3

Decimals for game 1:

3.6	0.03	13.1	29.6	11.9	0.7	33.7	0.04	21.9	0.125
10.1	0.42	0.07	2.9	0.29	19.5	5.52	23.1	9.6	1.8

Decimals for game 2:

0.023	0.9	5.74	0.245	7.3	1.2	23.8	16.5
0.068	8.7	0.12	1.8	0.78	42.7	0.2	3.6

[10]George Immerzeel and Melvin Thomas, eds., *Ideas from the Arithmetic Teacher: Grades 6–8* (Reston, VA: NCTM, 1982), p. 45.

Copyright © Houghton Mifflin Company. All rights reserved.

The Right Bucket Game Sheet for EXPLORATION 5.19

Turn	Decimals	Estimate	Reasoning	Actual product	Points
1					
2					
3					
4					
5					
6					
7					
8					
9					
10					
				Total	

Copyright © Houghton Mifflin Company. All rights reserved.

EXPLORATION 5.20 Operations: A Decimal Game

This game requires you to examine the effects of the four operations on decimal computation. As you move through the games, you should see improvement. At the end of each game, you will be asked to stop and reflect on your improvement.

Directions for playing the game

1. Make up four decimal numbers—for example, 3.6, 45.3, 1.23, 0.005.

2. From your list, select the three decimals that will make the answer to the given equation as large (or as small) as possible. Describe your strategy in words. For example, to make the equation

$$\square + \square - \square =$$

as large as possible using the decimals 3.6, 45.3, 1.23, and 0.005, you might try 45.3 + 3.6 − 0.005. Description of strategy: "I picked the two largest decimals to add and then picked the smallest decimal to subtract because I wanted to take away as little as possible."

3. If you are confident that you have the maximum number, go to Step 4. If not, try other combinations of numbers until you feel you have the combination that produces the largest number.

4. Make up four new decimal numbers. Give your strategy to another group. Have them use your strategy.
    Now check: Does this strategy produce the largest number?
    If yes, move on. If not, revise your strategy, and repeat Step 2.

1. *First Game*   Make the answer to the equation below as large as possible.

$$\square \times \square + \square =$$

2. *Second Game*   Make the answer to the equation below as *small* as possible.

$$\square \times \square \div \square =$$

3. Would your strategy in the second game (Step 2) change if we added parentheses as shown below?

$$\square \times (\square \div \square) =$$

Describe your hypothesis before doing any computations with the calculator. Then do some computations and decide whether to keep your original strategy or to revise it.

4. *Third Game*   This time you pick the operations, write the equation, and decide whether to make the answer to the equation as large or as small as possible.

Looking Back on Exploration 5.20

1. Does your strategy depend on the numbers chosen? For example, for the sample exploration, we would pick the two largest numbers to put in the first two boxes and the smallest number for the third box, regardless of the numbers. Is there a simple, general strategy for every game?

Copyright © Houghton Mifflin Company. All rights reserved.

EXPLORATION 5.21   Target: A Decimal Game

Materials

• Game sheet on page 129

Directions for playing the game

Note: These directions are written using an example that involves the operation of multiplication. The game can also be played using any of the other three operations: addition, subtraction, or division.

1. Select a number, an operation, a goal, and a winning zone; for example:

    Starting number   145
    Operation   multiplication
    Goal   1
    Winning zone   0.9 to 1.1 (or 0.99 to 1.01, or 1 to 1.1)

2. The first player selects a number to multiply the starting number by, trying to get the product of the two numbers in the winning zone (in this case, between 0.9 and 1.1).

3. If the product is not in the winning zone, then the product becomes the starting number for the second player.

4. Play continues in this manner until the product is within the winning zone.

1. Use the game sheet on page 129. Play the game several times with a partner. Record your game in the table provided on the game sheet. For example, the first turn in the game described in the directions might look like this:

Turn	Computation	Product	Reasoning
1	145 × 0.006	0.87	I knew that 100 × 0.01 would be 1. Because 145 is bigger than 100, I picked a number smaller than 0.01.

2. Describe one mental math strategy that you learned during this game.

Copyright © Houghton Mifflin Company. All rights reserved.

EXPLORATION 6.3   Unit Pricing and Buying Generic

Unit pricing is now common. Underneath most items in grocery stores is the unit price of the item. For example, if a 24-ounce jar of pickles costs $2.79, the unit price is 11.6¢, which means that the pickles cost 11.6¢ per ounce. Most states now require grocery stores to show the unit price below each item. One of the reasons for this law is that people tend to believe that larger items are proportionally cheaper. For example, they believe that a 64-ounce box of detergent will cost less than twice as much as a 32-ounce box. However, many companies use this belief to their advantage.

1. Let's say Paul Bunyan pancake mix comes in two sizes: 30 ounces and 50 ounces. The smaller box costs $2.89, and the larger box costs $4.69.

   a. Without using a calculator, determine which is the better buy.
   b. Use a calculator and record your process.
   c. Jackie did the problem this way: $\dfrac{30}{2.89} = 10.38$ $\dfrac{50}{4.69} = 10.66$

      What do 10.38 and 10.66 mean?
   d. What do you think of Jackie's method? Justify your response.
   e. Describe a situation in which you might buy the smaller item even if it cost more proportionally than the larger item.

An issue related to unit pricing is generic products. When I was growing up in the 1950s and 1960s, consumers could choose among the various name brands. Now, however, consumers can choose among the name brands or choose the generic alternative.

1. Are there products for which you are more likely to buy the generic brand or are not likely to buy the generic brand? What reasons do you give for choosing to buy the generic brand or choosing to buy the name brand?

2. One of the obvious reasons for buying the generic brand is that it is cheaper. Let's say you go to a store, and there you see a dispenser of Scotch brand tape that sells for $1.69 and a generic alternative that sells for $1.29. Both rolls contain the same amount of tape.

   a. How could we compare the two prices? Write down your thoughts.
   b. After hearing other strategies, describe and critique different ways in which we could compare the two prices.

3. Let's extend this question.

   a. Select a drugstore and collect and compare data on a name-brand item and a generic brand (or compare newspaper ads). Suppose you were writing an advertisement for the drugstore and you wanted to convince shoppers that they could save a lot of money by buying the generic item. Write the ad.
   b. Gather data on different sizes. Do you always get more for your money with the bigger size?
   c. Gather data from two different stores. How much cheaper is one product at one store than the other?

4. Let's say an average family of four decided to "buy generic" whenever possible. Over the course of a year, how much money would they save?

Copyright © Houghton Mifflin Company. All rights reserved.

EXPLORATION 6.4   Proportional Reasoning and Functions

In this exploration, we will explore a variety of functions that rely on proportional reasoning.

1. Solve each of the following problems.

    a. It costs Rita 50¢ for the first 3 minutes of a long-distance call to her boyfriend and 20¢ for each additional minute. If Rita calls her boyfriend and talks for 12 minutes, how much does the call cost?

    b. At Sam's Submarine Sandwich Shop, the cost of your submarine sandwich is determined by the length of the sandwich. If Sam charges $2 per foot, how much will a 20-inch sandwich cost?

    c. The first-grade class is measuring the length of a dinosaur by using students' footsteps. The first dinosaur is 6 of Alisha's footsteps or 9 of Carlo's footsteps. If the second dinosaur is 8 of Alisha's footsteps, how many of Carlo's footsteps will it be?

    d. A farmer has determined that he has enough hay for 4 cows for 3 weeks. If the farmer suddenly obtains 2 more cows, how long can he expect the hay to last?

    e. Certain bacteria can double in number in 1 hour. If we start with 1 bacterium, how many bacteria will there be after 20 hours?

    f. Consider the set of quadrinumbers below. How many dots does the fifth quadrinumber contain?

2. Discuss with your partner(s) solutions and strategies for each of the six problems in Step 1.

3. Determine a general formula for each problem below. Then discuss solutions and strategies for each problem.

    a. If it costs 50 cents for the first 3 minutes and x cents for each additional minute, and a person talks for y minutes, what is the cost?

    b. If the sandwich costs x dollars per foot and it is y inches long, what is the cost?

    c. If the first dinosaur is 6 of Alisha's footsteps or 9 of Carlo's footsteps, and another dinosaur is x of Alisha's footsteps, how many of Carlo's footsteps will it be?

    d. If there is enough hay for 4 cows for 3 weeks, how long would the hay last if there were x cows?

    e. If bacteria double in number every hour, how many bacteria will there be after x hours? Start with 1 bacterium.

    f. How many dots are in the xth quadrinumber?

4. For each problem, construct a graph that shows the relationship between the variables. Describe each graph as though you were talking to someone on the phone.

5. Describe the similarities and differences you see among these six problems. For example, the graphs of some families are straight lines and the graphs of others are not.

Copyright © Houghton Mifflin Company. All rights reserved.

SECTION **6.2**   **Exploring Percents**

Percents are a powerful tool that enables us to compare amounts and to describe change.

EXPLORATION 6.5   Percents

Sales

John sees that the local department store is having a sale. He goes to the store and finds that all televisions are 25% off.

1. Describe in words what that means.

2. Let's say he is interested in a particular television that normally sells for $400. If it is priced at 25% off, how much will he pay for it?

3. There are two common ways in which students solve this problem.

   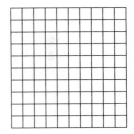

   - Ann: 25% of $400 is $100. $400 − $100 = $300. John pays $300.

   - Bela: 75% of $400 is $300. John pays $300.

   Ann doesn't understand what Bela did. How can you help her? You may, if you wish, use a grid like the one at the right.

4. Let's say that Joe gets a 5% raise and that he presently makes $8.00 per hour. One way to determine his new wage is to find 5% of 8.00 and then add that to 8.00. Using the ideas generated above, can you figure out how to determine his new wage with only one calculation? Describe the method.

Percent Decrease and Increase

Joshua is confused. He works for the Adamson Printing Company. Last year the economy was in such bad shape that all employees agreed to take a 20% cut in pay. However, this year the economy had improved so much that the company agreed to give everyone a 20% raise. Before the pay cut, Joshua was making $30,000 a year.

5. Explain why the 20% raise does not "undo" the 20% pay cut.

6. What raise would undo the 20% pay cut?

7. Determine a general formula that will tell you what percent increase will undo an *x* percent decrease.

Changes in Rates

In New Hampshire, where I live, there was a tremendous amount of controversy over the Seabrook nuclear power plant. The construction of the plant was held up many times. Eventually the company that made the reactor went bankrupt. It was bought out

Copyright © Houghton Mifflin Company. All rights reserved.

by another utility company. Part of the deal made to ensure that the new company would make a profit was that it could increase rates by *at least* 5.5% each year for 7 years.

8. Bill and Betty Olsen figured that their average monthly electric bill last year was $83.21.

   a.   If they use, on average, the same amount of electricity over the next 7 years, and their bill increases exactly 5.5% each year, what can they expect their monthly utility bill to be at the end of 7 years?

   b.   Meet with your partner(s) and discuss answers and solutions. If you think you would change your method in order to do a similar problem, describe how and why the new method works.

   c.   Jarrad used the following method:

   $$83.21 \times 0.055 = 4.57655$$

   $$83.21 + 4.58 = 87.79$$

   $$87.79 \times 0.055 = \text{etc.}$$

   Do you think Jarrad's method is valid? Why or why not?
   If not, what suggestions would you give to Jarrad? Justify your suggestions.

9. Last year the Olsens' combined income was $42,310. If their income increases by 5.5% each year, what will their income be at the end of 7 years?

Copyright © Houghton Mifflin Company. All rights reserved.

8. Move 4 coins from the 6-inch point to the other points, so that the ruler still balances. Write the numbers. Explain why it still balances in numerical terms without adding all the numbers and dividing by 6.

9. Make another arrangement that balances but is not symmetric and explain why it still balances.

10. Without adding the numbers and dividing by the sum of the numbers, predict whether the ruler will balance for the pennies located at the following spots. Explain your reasoning. If it doesn't balance, change the location of one penny to make it balance.

    a. 4   4   7   9
    b. 5   5   5   9
    c. 2   2   2   11

Mean as Fair Share

11. Let's say we determine how many pencils each child has in his or her desk and we get the results shown at the right. That is, one child had 2 pencils, another 3, another 4, and so on. Give pencils from those who have more to those who have fewer until everyone has the same number of pencils. How many pencils does each child have now?

12. Turn to the second arrangement. Again redistribute the pencils so that every child has the same number.

13. What does redistribution have to do with mean?

Putting It Together

14. Imagine the exam scores of a class of 10 students. Make three very different distributions that all have a mean of 80.

PART 2: Mean, median, and mode

A class of 11 college students has been surveyed and asked how many drinks they had in the past week. Below are several sets of data, each for a group of 11 students.

1. In each case, determine the mean, median, and mode:
    a. 0, 0, 2, 3, 5, 6, 7, 15, 17, 20, 35
    b. 0, 0, 0, 0, 0, 0, 15, 20, 20, 25, 30
    c. 3, 5, 5, 5, 7, 9, 12, 13, 15, 15, 17
    d. 0, 3, 4, 4, 10, 10, 10, 15, 15, 17, 22
    e. What do you conclude, from these four sets of data, about what the mean, median, and mode tell you.

2. a. Create 10 numbers where the mean and median are both greater than 10 and the mode is less than 10.
    b. Create 10 numbers where the mean and mode are both greater than 10 and the median is less than 10.
    c. Create 10 numbers where the mode and median are both greater than 10 and the mean is less than 10.

3. Explain what you learned from questions 1 and 2.

4. The instructor has graded the exam and tells you that the mean is 78. The instructor has all the data and has done every analysis you can think of.
    a. Write down what comes to your mind when you hear only this number.
    b. What does this number not tell you?

Copyright © Houghton Mifflin Company. All rights reserved.

    c. You can ask for one additional piece of information. Your goal is to understand better how the class did overall. What additional information would you like—why?

    d. You can see one graph. What graph would you like? Why?

5. You have applied for a job in a far-away state, and lots of information is available on the Web. You find that the average salary for a new teacher in that state is $46,250.

    a. Write down what comes to your mind when you hear only this number.

    b. What does this number not tell you?

    c. If you could dig up one additional piece of information about salaries in that state, what would it be? Why?

    d. Your instructor has given you that information. What did it tell you? What did it not tell you?

    e. If you could dig up one additional piece of information, what would it be? Why?

    f. Your instructor has given you that information. What did it tell you? What did it not tell you?

6. You are going to write a newspaper story on the exercise habits of students at a high school whose enrollment is 842. The statistics have to be kept short. You have room for one graph and 2 or 3 sentences. What graph would you want and what information would you include in those 2 or 3 sentences. Justify your choice.

7. Describe the understanding of *mean* that you have gained from this exploration. Consider, for example, what it means, what it tells us, what it doesn't tell us, and the like.

Copyright © Houghton Mifflin Company. All rights reserved.

EXPLORATION 7.4 **Explorations for Gathering and Analyzing Data**

In most of the cases, you will be asked to describe how you minimized measurement variation. Therefore, take some time before you record your official data to make sure that your procedure is as reliable (consistent) as possible.

NUMBER 1: How long is your reaction time?

Materials

Rulers.

Measuring reaction time is important for driving safety and for many occupations, for example, jet pilots. While there are many high-tech ways to measure one's reaction time, here is a low-tech way.

Procedure: Students work in pairs. The first person holds a ruler upright. The second person positions his/her fingers on either side of the ruler a specified distance apart. Without warning the first person lets go of the ruler and the second person tries to grab the ruler as quickly as possible. By subtracting the number where the person grabbed the ruler from the original number, we get a measure of the person's reaction time.

1. Describe what you did to minimize measurement variation.

2. Gather data for each person until you are confident that the number you have is the "true" reaction time for that individual. Justify your decision about how you determined that number.

3. Analyze the data for the whole class. What is the average reaction time? What else can you conclude from the data?

4. What did you learn?

NUMBER 2: How does more materials make the bridge stronger?

Materials

Spaghetti, Styrofoam cups, pennies.

Poke two holes on either side of the cup. Insert one strand of spaghetti through the holes and suspend the cup, either by one person gently holding the spaghetti on either side of the cup or by having two stacks of books and then suspending the spaghetti bridge between the two stacks.

1. Describe what you did to minimize measurement variation.

2. Gently drop pennies into the cup until the bridge breaks. Repeat the process until you are confident that you have minimized the measurement variation.

3. Repeat the process with a bridge of two strands, three strands, etc.

4. Analyze your results.

5. What did you learn?

NUMBER 3: How does weight affect the length of the jump?

Materials

Rubber bands, paperclips, weights (20 heavy washers work nicely), ruler, plastic bags.

Assemble your bungee jump.

1. Describe what you did to minimize measurement variation.

2. Place one weight in the bag and record how much the bungee cord increased in length.

Copyright © Houghton Mifflin Company. All rights reserved.

3. Repeat this 5 times.

4. Predict the result for 10 weights and describe your reasoning.

5. Repeat the process five more times.

6. Compare your result to your predicton.

7. Predict the result for 20 weights and describe your reasoning.

8. Using the same equipment, repeat the experiment one or two days later!

9. What did you learn?

NUMBER 4: How many times can you snap your fingers?

Materials

A watch or timer that can record 30 seconds.

1. Discuss what counts as a "snap."

2. Have each person snap his or her fingers as quickly as possible for 30 seconds.

3. Analyze the data.

4. Let people practice.

5. Repeat the process.

6. Compare this set of data with the first set of data.

7. What did you learn?

NUMBER 5: How good are you at penny horseshoes?

Materials

Pennies and a noncarpeted floor.

   Stand a designated distance from the wall and toss a penny toward the wall. The goals is to have the penny land as close as possible to the wall.

1. Describe what you did to minimize measurement variation.

2. Have each person do this a determined number of times.

3. Collect all the data.

4. Analyze the results.

5. What did you learn?

Copyright © Houghton Mifflin Company. All rights reserved.

EXPLORATION 7.5 How Many Drops of Water Will a Penny Hold?

Materials

Pennies, eyedroppers, paper towels.

Gently drop one drop of water at a time onto a penny until the water spills.

1. Spend some time thinking about how to minimize measurement variation. That is, ideally, you should be able to get the same number of drops each time. Describe what you did to minimize measurement variation.

2. Describe your procedure precisely, so someone reading your description could replicate what you did.

3. Do the experiment at least 20 times. Did the variation of the data decrease as you gathered more data? That is, did you become better at the process?

4. Analyze and present your results. Include:

   a. what you learned about minimizing measurement variation
   b. the analyses you did of your data: computations and graphs
   c. your conclusions

5. If possible, repeat this experiment with quarters. First, based on your penny data, predict how many drops a quarter will hold.

6. What did you learn from this experiment?

Copyright © Houghton Mifflin Company. All rights reserved.

EXPLORATION 7.6 How Accurate Can You Get the Whirlybird to Be?

Materials

Whirlybirds and measuring sticks or measuring tape. Make your own whirlybird. You can get the diagram from this website: http://www.pbs.org/teachersource/mathline/lessonplans/pdf/esmp/whirlybird.pdf

Hold the whirlybird about three or four feet above the ground.
Mark the point directly below your hand.
Drop the whirlybird and record how far it fell from that point.

1. Spend some time thinking about how to minimize measurement variation. Describe what you did to minimize measurement variation.

2. Describe your procedure precisely, so someone reading your description could replicate what you did.

3. Repeat this 10 to 20 times. Did the variation of the data decrease as you gathered more data? That is, did you become better at the process?

4. Analyze and present your results. Include:

    a. what you learned about minimizing measurement variation
    b. the analyses you did of your data: computations and graphs
    c. your conclusions

5. What did you learn from this experiment?

Copyright © Houghton Mifflin Company. All rights reserved.

EXPLORATION 7.7 Exploring Relationships Among Body Ratios

Materials

Measuring tape, string, rulers.

Having students gather and analyze data about various measurements of their bodies is an exploration that has appeared in many, many publications. You can see them on the website. This exploration has wonderful real-world applications, too. Forensic examiners and palentologists can give a very good approximation of the height of a person just from knowing the length of certain bones.

1. Using measuring tape, measuring sticks, and string, gather as much of the following data from each person in centimeters as your instructor directs.

A	Height			
B	Floor to belly button			
C	Belly button to top of head (A − B)			
D	Arm span			
E	Head circumference			
F	Radius			
G	Humerus			
H	Femur			
I	Tibia			
J	Foot			
K	Wrist			
L	Thumb			
M	Neck			
N	?			

Radius: Measure from the elbow bone to the wrist bone.
Humerus: Measure from the elbow bone to the top of the shoulder.
Femur: Measure from the top of the hip bone to the middle of the knee cap.
Tibia: Measure from the middle of the kneecap to the ankle bone.
Palm span: Holding the palm gently closed, measure the distance across the palm.
Hand span: Holding the fingers as far apart as possible, measure the distance from the tip of the little finger to the tip of the thumb.

2. a. Select one set of measurements (e.g., height, head circumference). Just from looking at the data, what do you see? What do you think the line plot or histogram for that data will look like?

   b. Analyze the data. Summarize what you found about that population: measures of central tendency, shape of the data, variation, etc.

Copyright © Houghton Mifflin Company. All rights reserved.

3. **a.** Determine those ratios below selected by your instructor. You can go to the course website to see how to use a spreadsheet to make this task less tedious!

    A:B   A:D   A:E   A:F   A:G   A:H   A:I   A:J

    C:B   K:L   M:K

    b. Analyze your data.

    c. Which ratios show the least variation?

4. Use the following formulas to determine each person's predicted height from your data.

    a. Determine the percent error in each case.

    b. Which bone seems to work best?

Male Height	Female Height
$2.9 \times$ length of humerus + 70.6	$2.8 \times$ length of humerus + 71.5
$3.3 \times$ length of radius + 86.0	$3.3 \times$ length of radius + 81.2
$1.9 \times$ length of femur + 81.3	$1.9 \times$ length of femur + 72.8
$2.4 \times$ length of tibia + 78.7	$2.4 \times$ length of tibia + 74.8

5. **a.** Make a scatterplot for as many of the pairs of variables from 2(a) as your instructor directs. You can go to the course website to see how to use a spreadsheet to make this task less tedious!

    b. Which scatterplots show the highest positive correlation?

6. In *Gulliver's Travels,* the Lilliputians had to make clothes for Gulliver. In the book, it is said that the Lilliputians only needed to measure Gulliver's thumb circumference in order to get his wrist, neck, and waist circumference, because each measure was twice the preceding measure. How true does that generalization hold for this class?

7. A crime has been committed and forensics people have recovered a handprint from the suspected culprit. If the length of the middle finger on the handprint is 8.9 centimeters, and they are pretty sure the suspect is an adult female, how tall do you think the suspect is?

8. Archaeologists have unearthed a statue of an adult female from ancient Greece. Unfortunately, both of the arms are missing. If the statue is 8 feet 6 inches tall, how long should the arms be.

9. Jack climbed the beanstalk and fell into a footprint of a giant. If the giant's footprint is 44 inches long, how tall is the giant?

10. A forensics team has unearthed a partial skeleton of an adult female from ancient Egypt. The length of the tibia bone (which goes from the middle of the kneecap to the ankle bone) is 36.5 cm. How tall was the person?

11. What did you learn from this Exploration?

Copyright © Houghton Mifflin Company. All rights reserved.

EXPLORATION 7.11 Designing and Conducting a Survey

In this exploration, you will design, conduct, and present results of a survey. My students report that this is both an interesting and a challenging project. In the first edition of the textbook, I did not have this exploration, because I knew that within the time constraints of this course, it is virtually impossible to determine a truly random sample. However, when I tried surveys with my class, I found that one of the major learnings from the exploration was an appreciation of the complexities of doing a survey and a more critical eye when reading and listening to the results of surveys. Because we are presented with survey results daily, in newspapers and on television, and because doing surveys is becoming increasingly common in elementary school textbooks, I decided to include a survey exploration.

1. Select a theme and at least four questions related to that theme. Carefully consider the questions that you will ask. You need to make sure that the respondents are answering the question you think you are asking! For example, when you ask whether they like the college, people define *like* in different ways. Similarly, you need to consider whether you want the question to be open-ended or forced-choice. For example, one or more groups in my class always collect data about alcohol use. If you ask how much alcohol someone has consumed in the past week, and some respondents write, "more than 10 drinks," this presents problems when you try to determine the average.

    You also need to determine whether you will ask respondents to fill out a questionaire or to respond verbally. *Note:* When you ask questions that are sensitive, such as questions about alcohol consumption, you are more likely to get honest responses if the respondents feel some sense of anonymity. For example, you can have them fill out the survey, fold the paper, and place it into a box with other surveys.

    Essentially, you want to minimize the many problems that may occur in the data, (such as missing data when someone completes only some questions and "bad" data when someone gives a response that can't be used or quantified).

2. Determine your target population. In most cases, your target population will be students at your college.

3. Devise, describe, and justify a strategy to get a representative sample. This discussion involves both logistics and justifications.
    a. Logistics: Where, when, and how will you collect the data?
    b. Justification: Why do you think the where, when, and how will give you a representative sample?

4. Determine how you and your group members will code the data. For example, if you want to compare males and females, then you have to code their responses separately.

5. Collect the data.

6. Analyze the data.
    a. First, analyze the data from your own sample. Determine the centers and variation for each question, and make sketches of appropriate graphs for each question.
    b. Compare results with other members of your group. Are your results (such as centers and variation) close or not?

7. Combine the data that you and your group members have collected. Determine what you have found and how best to present your findings to the class: centers and spreads, graphs, the text of your presentation, and the like.

8. Present your report to the class. Your instructor will specify the format for your report.

Copyright © Houghton Mifflin Company. All rights reserved.

SECTION 7.3 Exploring Concepts Related to Chance

Chances are, you have some understanding of probability concepts. For instance, you know that when you roll two dice, the probability of getting a 2 is less than the probability of getting a 7. You know that the probability of snow is less in Atlanta than in Buffalo.

Like the concept of "average," "chance" also has a specific meaning that you will explore in this section. In the process, you will use some tools you have developed in earlier chapters and will develop some new ones.

EXPLORATION 7.12 Heads and Tails and Probability

The simple question of flipping a fair coin has captured the attention of people for many years. While we know that theoretically the probability of heads or tails is 50%, you have probably noticed that sometimes we see heads or tails appear several times in a row. This exploration will deepen your understanding of the concept of randomness. For many people randomness and haphazard are synonyms. However, as you will soon discover, while we cannot predict the outcome of a particular tossing of a coin, we can fairly accurately predict the outcome of a large number of cases. That is, there are patterns in randomness!

1. Flip a coin 10 times and record the numer of heads.
2. Gather data from the entire class so that you have 50 samples of sample size 10.
3. Make a bar graph for 50 samples.
    a. How often did you get exactly 5 heads and 5 tails?
    b. How would you describe the shape of the bar graph?
    c. What generalizations can you make from the bar graph?
4. Make at least two more bar graphs, each consisting of 50 more samples. These graphs represent the same phenomenon, but the graphs are not the same.
    a. Why not?
    b. What generalizations can you make from observing these graphs?
5. Either by hand or with technology (using graphing calculators or the Excel program on the website), construct an ongoing line graph of the cumulative probability of heads.
6. How would you describe what happens as you get more and more data?
7. Summarize what you learned from this exploration.

Strings

1. Let's say we tossed a coin 50 times. On average, what do you think will be the longest string of heads or tails when throwing a fair coin 50 times?
2. Gather sufficient samples to answer this question. If your instructor directs you, use the website or other technology.
3. Make a bar graph from your data.
4. What is your response to this question? Support your conclusion.

Copyright © Houghton Mifflin Company. All rights reserved.

EXPLORATION 7.13 What Is the Probability of Having the Same Number of Boys and Girls?

In the children's story *Twenty-One Balloons,* by William Dubois, a man selects 25 families to join him on a remote, uninhabited island. However, there is a condition: He will only take families that have exactly 1 girl and 1 boy.

PART 1: Determining the probabilities experimentally

We will begin the exploration with a slightly more complex question.

1. What if a couple decided to have 4 children? What is the probability that they will have exactly 2 girls and 2 boys? Write your initial prediction and reasoning.

2. After the class discussion, revise your initial prediction and reasoning if you wish.

3. Simulate this problem. Collect samples of sample size 10. That is, how many times out of 10 do you get 2 boys and 2 girls? Do this a total of 50 times.

4. Make a line plot or bar graph of your data.

5. What do you think now?

6. How would you describe the distribution of the line plot or bar graph?

7. Using technology (graphing calculator or the Excel program on the website), construct an ongoing line graph of the cumulative occurrence of 2 boys and 2 girls.

8. How would you describe what happens as you get more and more data?

9. What is the empirical probability of having 2 girls and 2 boys in a family of 4?

PART 2: Determining the probabilities theoretically

Now we will develop the theoretical probability and see some neat patterns!

1. Detemine the theoretical probabilities for the following:
   a. For a family of 2: 2 girls, 1 girls, 0 girls.
   b. For a family of 3: 3 girls, 2 girls, 1 girl, 0 girls.
   c. For a family of 4: 4 girls, 3 girls, 2 girls, 1 girl, 0 girls.
   d. For a family of 5: 5 girls, 4 girls, 3 girls, 2 girls, 1 girl, 0 girls.

2. Represent the result in a table like the one below. The second row has been filled out. That is, if you have two children, there is one way to have 0 girls, there are two ways to have 1 girl, and there is one way to have 2 girls, for a total of four possibilities. Describe patterns you see in this table.

		Number of girls						
		0	1	2	3	4	5	Total
Number of children	1							
	2	1	2	1				4
	3							
	4							
	5							

3. Use the table to determine the theoretical probability of having 3 girls and 3 boys in a family of 6.

Copyright © Houghton Mifflin Company. All rights reserved.

EXPLORATION 7.14 What Is the Probability of Rolling Three Doubles in a Row?

Almost everyone in the class has either played *Monopoly* or heard of it. In this game, if you roll doubles, you can roll again. However, if you roll three doubles in a row, you go to jail.

1. What do you think is the probability of rolling three doubles in a row?

   Let us work up to answering this question. We will assume that we have fair dice. Therefore, the probability of each number is equal.

2. What if we roll two dice? What is the probability of rolling doubles? Write down your initial answer to this question and your reasoning.

3. In many classes, we have two different answers. All students agree that there are six ways of rolling doubles. However, some students believe that the sample space consists of 21 outcomes and some believe the sample space consists of 36 outcomes. The former group considers (1, 2) and (2, 1) to be the same outcome and the latter group considers them to be two different outcomes. If this is an issue in your class, discuss this question until you come to resolution. There are two ways, both having merit, to resolving this question.

   One involves a discussion of theoretical probability, that is, theoretically, what is the size of the sample space.

   The other involves collecting data. If the first answer is correct, then, with more and more data, the probability will tend to converge to 1/7 while if the second answer is correct, with more and more data, the probability will tend to converge to 1/6.

   Use either or both of these means to resolve the dilemma. If you originally had the wrong answer, take some time now to write what your thinking was that led you to the wrong answer and describe what helped you to understand the right answer.

4. Now we move up one level in complexity. What is the probability of rolling two doubles in a row? Write your first thoughts and your justification of your thinking. How confident are you that your reasoning is valid? As before, discuss this question with the class. Again, you can gather empirical data (see the website about how to do this with technology).

5. Now we move to the original question: What is the probability of rolling three doubles in a row? Write your answer and your justification.

6. If you rolled three dice and added the numbers, what is the probability that you would get 13?

Copyright © Houghton Mifflin Company. All rights reserved.

EXPLORATION 7.15   What's in the Bag?

Your instructor will have placed a large number of objects in a bag (e.g., different color marbles). There are many of these in the bag, and there are two colors. By sampling, you are to determine the proportion (percentage) of each.

1. Take out 10 objects and record the number of the designated color.

2. Once you get 10 samples of sample size 10, make a line plot.
    Looking only at the distribution, what do you see?

3. Collect more data until you have 50 samples of sample size 10.

    Make a line plot for these data.
    Looking only at the distribution, what do you see?

4. Use the simulator from the website or another source to generate four line plots, each containing 50 samples of sample size of 10.
    Describe similarities and differences between the various distributions.

5. Summarize what you have learned from this exploration about randomness, about patterns in randomness, and about what is the same and what is different when you look at different sets of data from the same bag.

Assessment:

6. Students drew samples from a bag that had 1000 marbles. The bag had blue and red marbles. Each of four groups collected 50 samples of sample size 10. On the next page are the line plots from each group showing the number of blue marbles in each sample of 10. Just by looking at the four line plots, state what you think is the proportion of each color and justify your results.

Copyright © Houghton Mifflin Company. All rights reserved.

Graph A
```
                              X
                              X
                              X
                              X
                       X      X
                       X      X
                       X      X
                       X      X
                       X      X
                       X      X
                X      X      X      X
                X      X      X      X
                X      X      X      X
        X       X      X      X      X      X
        X       X      X      X      X      X
        X   X   X      X      X      X   X
  0   1   2   3   4   5   6   7   8   9   10
```
        Graph A

Graph B
```
                                          X
                                          X
                                          X
                                          X      X
                                          X      X
                                          X      X
                                          X      X
                              X           X      X
                              X           X      X
                              X           X      X
                              X    X      X      X      X
                              X    X      X      X      X
                              X    X      X      X      X
                        X     X    X      X      X      X
                        X     X    X      X      X      X
  0   1   2   3   4   5   6   7   8   9   10
```
        Graph B

Graph C
```
                              X
                              X
                       X      X
                X      X      X
                X      X      X      X
                X      X      X      X
                X      X      X      X
         X      X      X      X      X
         X      X      X      X      X      X
         X      X      X      X      X      X
         X      X      X      X      X      X
  0   1   2   3   4   5   6   7   8   9   10
```
        Graph C

Graph D
```
                              X
                        X     X
                        X     X     X
                        X     X     X
                        X     X     X      X
                        X     X     X      X
                        X     X     X      X
                        X     X     X      X
                        X     X     X      X
                        X     X     X      X      X
                 X      X     X     X      X      X
                 X      X     X     X      X      X
  0   1   2   3   4   5   6   7   8   9   10
```
        Graph D

Copyright © Houghton Mifflin Company. All rights reserved.

EXPLORATION 7.16 How Many Boxes Will You Probably Have to Buy?

Cereal companies often place prizes inside boxes of cereal to attract customers. One of these promotions lends itself nicely to exploration. Many elementary school teachers have done this exploration, in a simpler form, with their students. It certainly captures their attention!

Inside specially marked packages of Sugar Sugar cereal, you can get a photograph of one of six sports figures. Many kids will want to have one of each.

1. If we want to get all six prizes, how many boxes do you predict we will have to buy, on average?
    a. Write your prediction and justify your reasoning.
    b. (optional) Gather everyone's prediction and summarize what the class prediction is—average and variation, distribution.

2. Each group simulates the situation 10 times.

3. a. On the basis of 10 simulations, what do you predict now? Justify your reasoning.
    b. What have you learned or noticed from these 10 turns?

4. Collect the data from the whole class.

    a. Analyze those data. What do you predict now?
    b. Indicate which operations on the data were useful—computations of mean, median, mode, range, standard deviation, etc. Which graphs were useful?

Note: You can go to the website and download an Excel program to simulate this situation for different numbers of prizes.

Copyright © Houghton Mifflin Company. All rights reserved.

EXPLORATION 7.17 More Simulations

The following situations present real-life questions for which determining the theoretical probability is either tedious or impossible. Thus they lend themselves to simulations in which we can determine the experimental probability.

For each of the following questions, develop and execute a simulation plan to help you answer the question. Your instructor will specify the format for your reports.

1. *The basketball game is on the line*   Horace has just been fouled, and there is no time on the clock. His team is down by one point, and he gets two shots. Horace is an 80% free-throw shooter. What percent chance of winning in regulation time does his team have?

2. *Genetics*   Maggie and Tony have just discovered that they are carriers of a genetically transmitted disease and that they have a 25% chance of passing on this disease to their children. They had planned to have four children. If they do have four children, what is the probability that at least one of the children will get the disease?

3. *Overbooking*   Airlines commonly overbook; that is, they sell more tickets for a flight than there are seats.

   a. Why do you think airlines overbook?
   b. An airline has a plane with 40 seats. What information will help it to decide how much to overbook?
   c. Using the information from your instructor, develop a simulation plan to help the airline decide how many tickets to sell. Meet with another group to share ideas. Note any changes in your plan.
   d. Do the simulation. What is your conclusion? Support your conclusion.

4. *Having children*   In an attempt to deal with overpopulation, the Chinese government has a policy that couples may have only one child. One of the biggest problems with this policy is that in Chinese culture, having a boy is preferable. As a result of the policy and the preference for boys, infanticide is not uncommon in China (and in other countries too—for example, in parts of Nepal, where I served in the Peace Corps). That is, it is not uncommon for parents to kill their newborn baby if it is a girl. Obviously, there are many ways to deal with this problem.

   Two questions that arise from this policy are: (1) how would it affect family size, and (2) would it affect the ratio of boys to girls? Assume that this policy was implemented. Develop and run a simulation to answer these questions.

Copyright © Houghton Mifflin Company. All rights reserved.

CHAPTER

8 Geometry as Shape

At its root, geometry involves shape. In Chapters 9 and 10, we'll look at various applications of geometry and connections between geometry and other fields of mathematics. We will start with explorations that will orient you to a more geometric way of thinking.

Explorations 8.1, 8.2, and 8.3 make use of three types of manipulatives—Geoboards, tangrams, and polyominoes—that are helpful in developing a strong understanding of many important mathematical ideas. These explorations have multiple parts and address topics from throughout Chapter 8. Your instructor may choose to use parts of these explorations in combination with other explorations from the individual sections, which follow these opening explorations.

EXPLORATION 8.1 Geoboard Explorations

Geoboards are a popular manipulative in elementary schools and are very versatile. Smaller Geoboards generally contain 25 pegs (5 rows of 5 in a row), and larger Geoboards generally contain 100 pegs. There are also circular Geoboards, generally with 24 pegs arranged in a circle.

PART 1: Communication

This first part of the exploration both serves as an introduction to Geoboards and reinforces the need for clear communication when talking about shapes.

Instructions: In this exploration, you will work in groups of 3.

Persons A and B each have a Geoboard (or Geoboard Dot Paper) and sit back to back. Person C is the observer.

Person A makes a figure on the Geoboard. Next, using only words, person A gives directions so that person B can construct the same figure on his or her Geoboard. The observer watches without comment and will give feedback at the end. Person B's responsibility is to ask for clarification whenever person A's directions are not clear.

Copyright © Houghton Mifflin Company. All rights reserved.

1. After you are finished, compare Geoboards. The figures may or may not be identical. In either case, listen to person C's feedback and discuss how to make communication easier and pinpoint places where communication broke down.

2. Rotate roles: Person C now makes the figure and gives directions, person A makes the copy, and person B is the observer.

3. Repeat this process once more so that each person has a turn in each role.

4. Afterwards, describe what ideas or terms you learned to make communication easier.

PART 2: Challenges

You may use your Geoboards to solve the problems below. Afterwards, transfer your answer to Geoboard Dot Paper (at the back of this book).

Children often say, "I wonder if . . ."—for example, "I wonder if you can have a figure with exactly 3 right angles." "I wonder how many different figures you can have that have only one peg inside."

The challenges below address a few of the hundreds of possible "I wonder" questions. Pursuing these challenges can help you develop problem-solving and reasoning skills and also help you to understand some of the properties of geometric figures.

In Steps 1–3, construct the figures on your Geoboard (or Geoboard Dot Paper) or explain why you think it is impossible to do so.

1. **a.** A figure with just 1 right angle.
    b. A figure with 2 right angles.
    c. A figure with at least 1 right angle but no sides parallel to the edges of the Geoboard.
    d. A figure with 6 right angles.

2. A figure with exactly 2 congruent, adjacent sides.

3. Two figures with different shapes but the same area.

4. **a.** Find all possible squares that can be made on a 25-peg Geoboard.
    b. Make a quadrilateral with no parallel sides.
    c. Make a parallelogram with no sides parallel to the edge of the Geoboard.
    d. Make two shapes that have the same shape but are different sizes.

5. **a.** Make a pentagon. Then list all the attributes of this pentagon that you can.
    b. Make another pentagon and list all the attributes of this pentagon that you can.
    c. Now list the characteristics they have in common.

PART 3: Making triangles and quadrilaterals

Problem 1 appears in the "Navigating Through Geometry in Grades 3–5" book. You can examine the children's work after the exploration.

1. How many different right triangles can you make on a 5 × 5 Geoboard? There is a lot to think about here:

    • Can you think systematically, as opposed to just randomly?

    • How will you represent your triangles on the paper so you can keep track of them?

    • How do you know an angle is a right angle? There are right triangles whose bases are not parallel to the edges of the paper!

    On the paper you turn in, include

    a. Your method(s) for systematically determining all the triangles.
    b. A brief description of why you placed the triangles in the order you did.

Copyright © Houghton Mifflin Company. All rights reserved.

    c. How you identified right angles whose sides were not parallel to the edges of the paper. Here we are looking for more than "it looked like a right angle" or "I used a protractor."

Problem 2 was posed in the October 2000 issue of *Teaching Children Mathematics*. The solution, containing results submitted by elementary school teachers at various grade levels, appears in the October 2001 issue.

2. How many different triangles can you make on a 5 × 5 Geoboard that have no pegs in the interior of the triangle? Refer to Problem 1 for further discussion and hints. On the paper you turn in, include

    a. Your method(s) for systematically determining the triangles.
    b. A brief description of why you placed the triangles in the order you did.

3. How many different triangles can you make on a 5 × 5 Geoboard? Refer to Problem 1 for further discussion and hints. On the paper you turn in, include

    a. Your method(s) for systematically determining the triangles.
    b. A brief description of why you placed the triangles in the order you did.

Your instructor might choose to make the problem slightly smaller by specifying 4 × 4 instead of 5 × 5 Geoboards. This problem is substantially more complex than the previous two. You really need to think about being systematic, and you also need to think about how you will represent your answers. That is, it makes it easier to compare answers if there is some kind of organization of your triangles. Some students have found it useful to invent their own notation for triangles, such as 1 × 2, 1 × 3, etc.

PART 4: Recognizing and classifying figures with Geoboards

Geoboards can also be used to help students appreciate the need for names for various geometric figures and geometric ideas. In elementary, middle, and high school, you learned the names and properties of a number of shapes, such as isosceles triangle, parallelogram, and regular polygon. For example, there are (infinitely) many isosceles triangles, but regardless of their size and angles, all isosceles triangles have certain attributes in common. In this exploration, you explore different kinds of subsets from a small universe of geometric figures.

1. a. Cut out the set of quadrilaterals on page 179.
    b. Separate them into two or more groups so that the members of each group are alike in some way and so that each quadrilateral belongs to exactly one group.
    c. Describe each group so that someone who couldn't see your figures could get a good idea of what each group looked like.
    d. Name each group. If you know of a mathematical name for one or more groups, use it. If you don't know of a mathematical name, make up a descriptive name that fits the group.
    e. Record the figures in each group.

2. Repeat this process as many times as you can in the time given. That is, in what other ways can we separate these figures into two or more groups so that every group has one or more common characteristics?

3. After the class discussion, answer the following questions:

    a. What did you learn from this exploration?
    b. How do definitions help us to communicate when discussing geometry?

Copyright © Houghton Mifflin Company. All rights reserved.

Figures for EXPLORATION 8.1

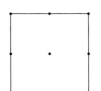
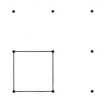
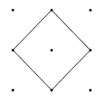

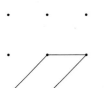

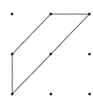

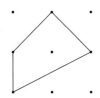
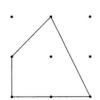

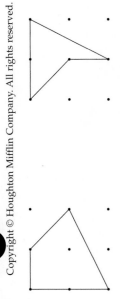

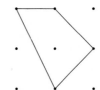

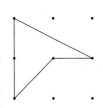

Copyright © Houghton Mifflin Company. All rights reserved.

EXPLORATION 8.2   Tangram Explorations

Like Geoboards, tangrams are very versatile manipulatives that we will use on several occasions in this and the next two chapters. Tangrams were invented in China at least two hundred years ago, but we are not sure by whom or what for. They quickly became a popular puzzle, because there are so many different things you can do with them! The English word *tangram* probably came from American sailors who referred to all things Chinese as Tang, from the Cantonese word for China.

You will use tangrams provided by your instructor or make your own set from the template at the back of the book.

PART 1: Observations, discoveries, and questions

The directions here are very simple: Play with the tangrams for a while. You may have questions for which you have some preliminary responses, and you may have some questions for which you don't have an initial response. As you explore the shapes, you will notice things: patterns and relationships among the various pieces. You might make interesting shapes. You may ask yourself various "what if" questions.

Record your observations: patterns, discoveries, conjectures, and questions.

PART 2: Puzzles

As you may have found, you can make a number of interesting shapes with tangrams. Below are some famous puzzles that call for careful thinking.

1. **a.** Use all seven tangram pieces to make each of the five figures shown on page 183. Sketch your solution.
    b. Describe any thinking tools that you became aware of while figuring out how to make the figures.
    c. Describe any new observations: patterns, discoveries, conjectures, and questions.

2. Make your own puzzle. Name it. Give it to someone else to solve.

3. Make a figure with the tangram pieces. Write directions for making that figure, as though you were talking on the phone to a friend. Exchange directions with a partner. Try to make the figure from your partner's description.

    a. Discuss any problems that either of you encountered in understanding what the other had written.
    b. Now make a new figure and write directions for making that figure, as though you were talking on the phone to a friend.

Copyright © Houghton Mifflin Company. All rights reserved.

Figures for EXPLORATION 8.2, Part 2: Puzzles

1.

Copyright © Houghton Mifflin Company. All rights reserved.

PART 5: Challenges

Use the tetromino and pentomino sets that you made in Part 2.

Note: In each of the challenges, describe strategies you used beyond guess–check–revise.

1. Using all five tetrominoes, completely fill grids (a) and (b) on page 191.

2. Using four different pentominoes, fill grids a and b on page 192.

3. Selecting among your set of pentominoes, solve the following puzzles.
    a. Use 4 different pentominoes to make a 4 × 5 rectangle.
    b. Use 6 different pentominoes to make a 3 × 10 rectangle.
    c. Use 5 different pentominoes to make a 5 × 5 square.

4. Using your whole set of 12 pentominoes, arrange the pentominoes so that the space enclosed by the pentominoes is as large as possible.

5. This challenge comes from a story, probably fiction, that goes like this: The son of William the Conqueror and the dauphin of France were playing a game of chess. At one point, the dauphin became angry and threw the chessboard at William's son. The board broke into thirteen pieces, 12 different pentominoes and 1 tetromino. Make a chessboard with the 12 different pentominoes and 1 tetromino.

6. Make up your own challenge. Describe your thinking process and a solution.

PART 6: Questions about categories

1. Juan sorted the pentominoes into these four groups. Write the directions for sorting so that the reader of this description will know how to put the pentominoes into the correct groups.

GROUP 1   GROUP 2   GROUP 3   GROUP 4

2. Lakela sorted the pentominoes into these three groups. Write the directions for sorting so that the reader of this description will know how to put the pentominoes into the correct groups.

GROUP 1   GROUP 2   GROUP 3

3. Ari sorted the pentominoes into these four groups. Write the directions for sorting so that the reader of this description will know how to put the pentominoes into the correct groups.

GROUP 1   GROUP 2   GROUP 3   GROUP 4

Copyright © Houghton Mifflin Company. All rights reserved.

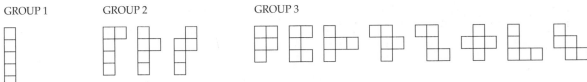
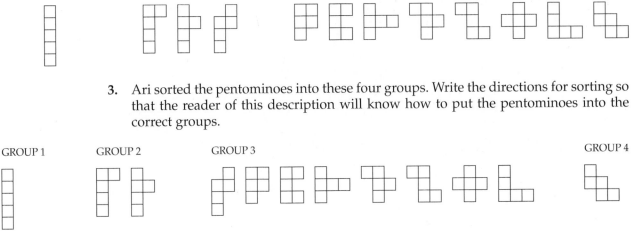

Hexomino Sheet

Copyright © Houghton Mifflin Company. All rights reserved.

SECTION **8.1** Exploring Basic Concepts of Geometry

In order to work effectively with geometric ideas, we need to have a foundation of postulates, definitions, and theorems. As you may recall from high school geometry, the Greek contribution to geometry and mathematics is enormous. Although this textbook does not attempt to replicate the kind of formal geometric work you did in high school, I do believe it is helpful for elementary teachers to explore concepts and ideas that students will further examine in high school in order to have a better sense of the connection between the explorations students do in elementary school and the geometric knowledge they need to bring to middle school and high school.

EXPLORATION 8.4   Manhole Covers

Geometry serves many purposes in our world. In this exploration, we will look at the geometric reason for the shape of one object.

PART 1: Manhole covers

1. Why do you think manhole covers are round? Write down what you think.

2. In order to understand why manhole covers are round, we will cut out a variety of shapes.

   a. Trace the circle and the square on pages 199 and 200 onto a blank sheet of paper, some space apart. Cut out the circle and the square in such a way that you only cut on the lines of the square and the circle. That is, fold the square in half and cut around the edges; do the same with the circle.

   b. You now have models of a circular and a square manhole cover. What do you notice?

   c. If you hold the square on edge and turn it, the length of the diagonal of the square is significantly longer than the length of one side, so the square can easily fall into the hole. The longest line segment you can draw across the hole made by the circle is equal to the diameter of the circle. How do you think the designers of manhole covers addressed this so that the circular manhole cover would not fall into the hole? Write your thoughts.

   d. How big a lip would you have to make for a square manhole cover so that the cover could not accidentally fall through?

   e. How big would the lip need to be for a pentagonal manhole cover? For a hexagonal manhole cover?

PART 2: Cutting out shapes with the least number of cuts.

1. Trace the square on page 199 to another sheet of paper. Then fold the paper so that the fold cuts the square in half. Now you can cut the paper beginning at A and making your way around the square. To cut the square out while keeping the sheet of paper intact, we made use of some properties of a square: We know that a square has symmetry, so if we fold it in half, the top half will fit exactly onto the bottom half. Thus we need only cut half of the square to cut the whole square out of the paper. Looking a little closer, we see that the properties of a square that make it easier to cut by folding are that the opposite sides of the square are equal (congruent) and that all four angles are also equal (congruent).

Copyright © Houghton Mifflin Company. All rights reserved.

What we just did counts as 3 cuts— from A to B, from B to C, and from C to D. There are two ways to fold the paper so that you can cut out the square with one cut. Try to figure how to cut the square with only one cut. Write down your solution. If you can only get it down to 2 cuts, write down your solution.

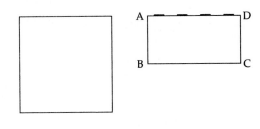

2.   Trace each of the polygons on the page to another sheet of paper. Figure out how to cut the shape out with the least number of cuts. Before you do so, think of your strategy and the properties of the shape that enable you to make fewer cuts than the number of sides. Once you have your solution:

a.   Describe your strategy so someone could read your description and be able to do it.

b.   Describe the properties of the figure that enabled you to solve the problem.

Copyright © Houghton Mifflin Company. All rights reserved.

Figures for **EXPLORATION 8.13**

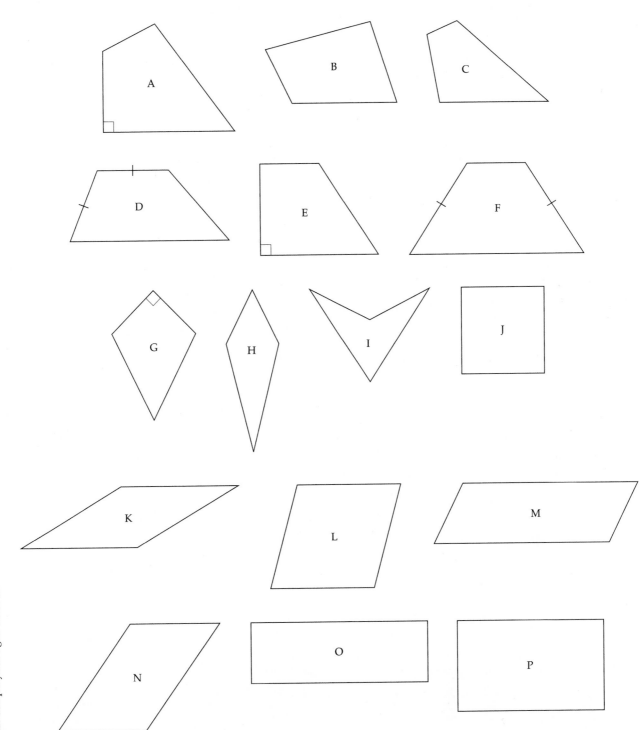

Copyright © Houghton Mifflin Company. All rights reserved.

SECTION **8.3**   **Exploring Three-Dimensional Figures**

When we make complex objects, whether they be cars or houses or sculptures, they have to be designed first, and the actual construction is done from blueprints or drawings— that is, two-dimensional representations of the object. More and more textbooks and other curriculum materials are emphasizing spatial visualization. It is very likely that you will do explorations like the ones below, at a simpler level, with your future students. The following explorations provide you an opportunity to work on these concepts and ideas and to do so in a way that should be both enjoyable and challenging.

EXPLORATION 8.14   Exploring Polyhedra

PART 1: Comparing polyhedra and polygons

Your instructor will direct you to a set of polyhedra—either physical models or pictures.

1. How are polyhedra like polygons and how are they different from polygons?

2. Focus now on the set of prisms. What attributes do they *all* have in common?

3. Focus now on the set of pyramids. What attributes do they *all* have in common?

4. How are prisms like pyramids and how are they different from pyramids?

5. In this step, you explore what terms used to describe polygons are also used to describe polyhedra, what new terms are needed, and why.
   a. If a polygon is a simple, closed curve, is there an analogous definition for polyhedra?
   b. All polygons have vertices, line segments, and angles. Do these terms work for describing polyhedra? Do we need new terms too? If so, why?
   c. With some polygons, we use the terms *base* and *height*. Do those terms apply to all polyhedra?
   d. We can call a polygon with *n* sides an *n*-gon. Is there an analogous notion for polyhedra (i.e, is a polyhedron with *n* sides an *n*-dron)? Why or why not?
   e. If there are regular polyhedra, how might they be defined?

6. When we want to describe one of the line segments of a polygon, we use the term *side*. However, when describing one of the line segments of a polyhedron, mathematicians use the term *edge*. Why do you think this term is used instead of *side*?

7. When describing polyhedra, many people use the term *corner*, yet this term is not found in mathematical dictionaries. Why do you think that is true? Define *corner* in your own words.

8. a. We speak of convex and concave polygons. Do those terms apply to polyhedra? Support your thinking.
   b. If you defined *convex* in Exploration 8.9, look at the definitions that were created. Do some of the definitions apply to three-dimensional figures better than others do?

PART 2: Determining congruence with polyhedra

You can read childrens' solutions to this problem in the April 2002 issue of *Teaching Children Mathematics*, pp. 444–445.

Copyright © Houghton Mifflin Company. All rights reserved.

1. Each group will make at least two different polyhedra from two squares and six equilateral triangles, where the lengths of the sides of the squares and of the equilateral triangles are equal.

2. Describe your thinking process for making the polyhedra and in going about making another one that was different from the first one you made.

3. Examine the class set. How many different polyhedra are there in the set? Here *different* means "not congruent."

    If two structures were close to congruent, what did you do to determine whether they were in fact congruent or not congruent?

4. Determine the number of edges, vertices, and faces for the different figures. You will use this information in Exploration 8.15.

PART 3: Sort polyhedra into groups

1. Take your set of polyhedra or the set given by your instructor.

2. Sort the polyhedra into two or more groups. Describe the criteria for each group. If your instructor gives you a new object, determine which of your groups it goes into. Discuss this until all students agree or until there is an impasse.

3. Repeat the process as many times as your instructor indicates.

PART 4: Writing directions

1. Write directions for making a structure that you have selected or a structure selected by your instructor.

2. Compare descriptions.

3. a. Give the directions to someone who is not in this class. Have this person try to make the polyhedron from your directions.

    b. If problems arose, describe and analyze the problem areas. What caused the confusion?

    c. Revise your directions, if necessary.

Copyright © Houghton Mifflin Company. All rights reserved.

3. Make a picture of a building that contains between 10 and 15 cubes and will be "challenging" to draw. For example, a building with fewer cubes in a back row than in a front row. Write the directions for making that building. Explain any difficulties and how you overcame them.

PART 4: Continuation

There is another method for giving directions for making block buildings. Rarely do students come up with this method in Part 2, and yet it is a method that is commonly presented to children in elementary and middle schools. One reason for this is that although it is not an obvious way, there is much potential for learning by examining this method. This method comes from the idea of taking a picture from all sides. For example, consider the building below.

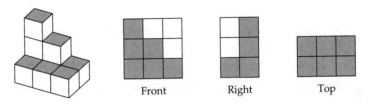

To the right of the building are three views of the building: the front view, the right side view, and the top view. To understand this method, some students find it more useful to imagine looking at the building from ground level. Envision crouching down so your eye is at the same level as the bottom of the building. Imagine seeing only the shadow of the building, like seeing someone's silhouette.

1. Make the buildings from the three views given in the exercises below. Afterwards, you will be asked to evaluate this method by comparing to the methods you developed earlier.

   a. Front   Right   Top

   b.

   c.

   d.

   e.

   f.

Copyright © Houghton Mifflin Company. All rights reserved.

2. Draw the front, right side, and top views of the buildings below.

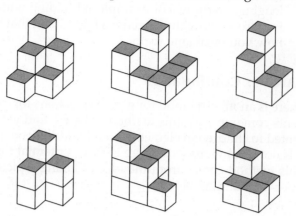

3. Describe the advantages and strengths of this method of giving directions for making a building. What kinds of buildings would it be most suited for, or is it an all-purpose method?

4. What do you think of this method compared to the ones you developed earlier?

PART 5: What if

1. What if you could make buildings with interlocking cubes—for example, the one at the right?

   a. Make a nonsimple building with interlocking cubes, and write directions for making the building using any of the methods already devised or a new method (which could be a modification of previous methods). Have another group make the building using your directions.
   b. Report the results. If it wasn't successful, explain the glitches and how you solved the glitches. Then repeat part (a).
   c. What did you learn from this?

2. What if the buildings were not restricted to cubes? Make a set of directions for one of the buildings below.

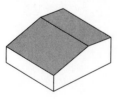  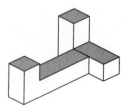

3. *Looking back:* What did you learn from this exploration?

Copyright © Houghton Mifflin Company. All rights reserved.

EXPLORATION 8.18   Cross Sections

One important spatial visualization skill is being able to imagine what happens when physical objects are changed or moved. Examples include being able to imagine how a new arrangement of furniture will work in a living room and being able to figure out whether all of a set of objects will fit into a given space (such as a suitcase or the trunk of a car).

Here, we will explore the notion of cross sections. Consider the cube at the right. If we sliced it in half (as though a knife made a vertical slice), what would the newly exposed face (that is, the cross section) look like? If we sliced it at an angle, what would the new face look like now?

PART 1: Predictions

For each figure below, first predict what the sliced face will look like. Then check your prediction either by making a clay figure and slicing it or by some other means. Describe any observations.

PART 2: Descriptions of intersections

In Part 1, you were given a picture representing a specific cross section of a solid object. What if you had no picture?

1. Describe an object and an intersecting plane, using only words.

2. Have another group describe what slice they would make, on the basis of your description. If they are able to decode your directions accurately, great. If there are problems, analyze them, and then revise the description to address the problems raised.

a. 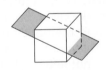   b. 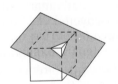   c.   d.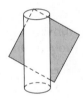

e.    f. 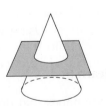   g.    h.

i.   j.   k.   l.

Copyright © Houghton Mifflin Company. All rights reserved.

EXPLORATION 8.19 Nets

This exploration continues our work in looking at connections between two-dimensional and three-dimensional objects. A net is simply a two-dimensional figure that will fold up into a three-dimensional figure. For example, the shape below at the left is a net for a cube, because it will fold up into a cube. The shape at the right below is not a net, because it will not fold into a cube or other three-dimensional figure.

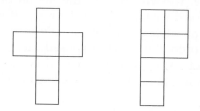

PART 1: Pentominoes and hexominoes that are nets

1. If you did Exploration 8.3, you found that there are 12 pentominoes. Use the table on p. 193 to predict which pentominoes are nets for a cube with no top.

   a. Predict and briefly explain the reasoning behind your prediction.
   b. Use the Polyomino Grid Paper to see whether it does fold up.
   c. If your prediction was correct, great. If not, why were you wrong? What did you learn?

2. It would be nice if all the pentominoes that are nets had one characteristic in common that would enable us to remember which pentominoes are nets. However, this is not always the case.

   a. Look at the pentominoes that are nets (those that do fold up) and see whether they all have one characteristic in common that none of the not-nets have. If not, you will do what mathematicians do when a general theorem is not possible or has not been discovered yet: You will make cases. For example, one such case is all nets that have four cubes in a row and then one on either side will fold up. Do you see that one? What about the other pentominoes that are nets?
   b. Now look at those pentominoes that are not nets. Is there one characteristic they all have in common that none of the nets have? If not, are there certain characteristics that preclude the possibility of folding into a cube?

3. If you did Exploration 8.3, you found that there are 35 hexominoes. Look at the table on page 195 showing all 35. Using what you learned about nets, name those hexominoes that you are close to 100% sure will fold up into a cube. Explain why. Then name those hexominoes that you are not close to 100% sure will not fold up into a cube. Explain why not.

4. Now look at the hexominoes that do fold up into a cube. Once again, is there one characteristic they all have in common that none of the not-nets have? If not, are there certain characteristics that some of the nets have in common, as we found for pentominoes?

Copyright © Houghton Mifflin Company. All rights reserved.

PART 2: Making nets

One of the most common nets that people (especially those who recycle) experience regularly is a flattened cereal box.

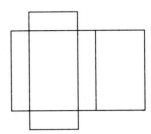

1. Here is one set for a standard cereal box.

    a. Give the dimensions.
    b. What do you notice about this net?
    c. What attributes and characteristics does it have?

2. Now make as many *different* nets as you can for the cereal box. If you use Polyomino Grid Paper, you can make the front and back 4×2 rectangles, the sides 4×1 rectangles and the top and bottom 2×1 rectangles.

3. Examine the whole class set of nets for a cereal box. What do you see? Are some of these more similar than others—that is, in the same family, just as a triangular prism, a cube, and a box are all in the same family that we call "prism"?

4. What attributes do all of the nets have in common?

5. Do you see any connections among the dimensions of each side?

6. Do you see any connections between the nets for cereal boxes and our work with pentominoes and hexominoes? If so, describe the connections.

7. Make as many nets as your instructor asks for the following shapes.

    a.   b.   c.
    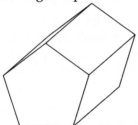   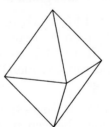   

    d. Pentagonal prism   e. Octahedron   f. Soccer ball
    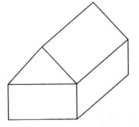   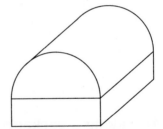   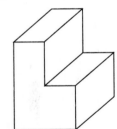

PART 3: Will the net fold up?

Now that you have experience making nets for shapes, let's go the other way. Following are some possible nets. Predict which will fold up into a polyhedra and which will not. Give your reasons. If you predict it will, sketch the polyhedron or describe it.

Copyright © Houghton Mifflin Company. All rights reserved.

PART 4: Making a package for an object

We see packages everywhere. Most packages are manufactured as flat surfaces that are then cut and folded and glued or stapled to become packages.

1. Select an object for which you will make the package. It might be similar to a package you have seen, and you want to improve or modify it, such as the package for french fries. It might be something not normally packaged, such as a bunch of bananas. You can consider large objects, in which case you would make a scale model package.

2. Think about the package—the shape, how much space you want, padding, whether the object should be seen, how the package will be opened and (if necessary) closed again.

3. Make the package. Turn in
    a. The package
    b. The net for your package
    c. Your explanation of why this package is "right" for this product—a sentence or two.

4. Describe how you used your spatial sense, estimating skills, and measuring skills in your design—a paragraph that essentially describes what mathematical knowledge and skills you used in this project.

Copyright © Houghton Mifflin Company. All rights reserved.

Figures for EXPLORATION 9.1, PART 1: Slides, flips, and turns

3. a.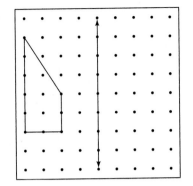

b.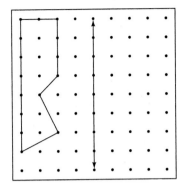

c.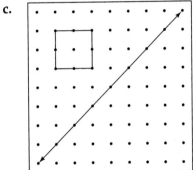

d.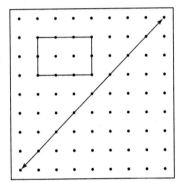

Copyright © Houghton Mifflin Company. All rights reserved.

Figures for **EXPLORATION 9.1, PART 2: Symmetry**

1. a. 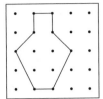   b.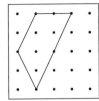

    c. 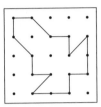   d.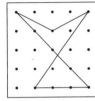

2. a. 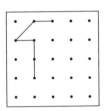   b.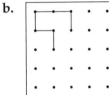

    c. 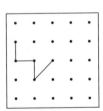   d.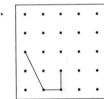

3. a. 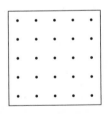   b.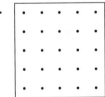

    c. 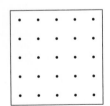   d.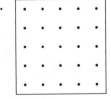

Copyright © Houghton Mifflin Company. All rights reserved.

Figures for EXPLORATION 9.1, PART 3: Similarity

a.
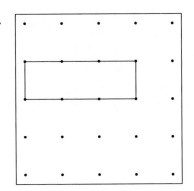

b.
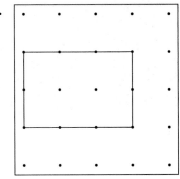

c.
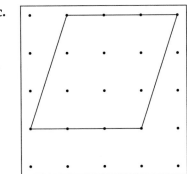

d.
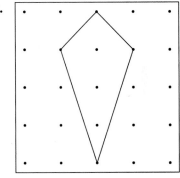

e.
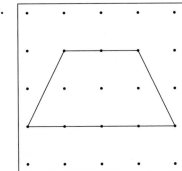

f.
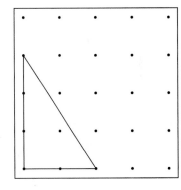

g.
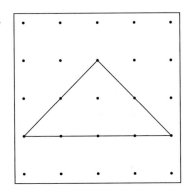

h.
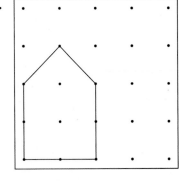

Copyright © Houghton Mifflin Company. All rights reserved.

EXPLORATION 9.2 Tangram Explorations

PART 1: Symmetry

1. Make different shapes from your tangram pieces so that each overall shape has one line of symmetry.

2. Make different shapes from your tangram pieces so that each overall shape has two lines of symmetry.

3. Make different shapes from your tangram pieces so that each overall shape can be rotated and still look the same. Specify the nature of the rotation symmetry of the shape.

PART 2: Similarity

Rectangles

1. a. With your group, make as many different rectangles as you can, using the pieces from one tangram set. Make a copy of each rectangle by tracing it.
    b. Compare your results with those of another group (or of the whole class). Add any new drawings of rectangles to your group's collection.

2. a. Separate the set of rectangles into subsets so that all the rectangles in a subset are similar to each other.
    b. Describe how you determined which rectangles are similar. Then write a definition or a rule for *similar* rectangles. This will be your first draft.
    c. Share your definition with other groups.
    d. If your definition or rule has changed, note the change and explain why you like this version better than your original version.

Trapezoids

3. a. Make as many different trapezoids as you can, using pieces from one tangram set. Make a copy of each trapezoid by tracing it.
    b. Compare your results with those of another group (or of the whole class). Add any new drawings of trapezoids to your group's collection.

4. a. Separate the trapezoids into subsets so that all the trapezoids in a subset are similar to each other.
    b. Use your definition or rule from Step 2 to determine which trapezoids are similar. If it still works, move on. If it doesn't, explain why, and then modify your definition or rule.
    c. Share your definition with other groups.
    d. If your definition or rule has changed, note the change and explain why you like this version better than your original version.

Extensions

5. Can you extend your definition of similarity to any geometric figure? Make a tangram figure that is not a triangle or a quadrilateral. Now make a similar figure on another sheet of paper. Explain how you did it and justify your solution.

6. Irma says that all of the triangles that you can make with tangrams will be similar to one another. What do you think of Irma's conjecture? Explain your reasoning.

Copyright © Houghton Mifflin Company. All rights reserved.

EXPLORATION 9.3 Polyomino Explorations

PART 1: Slides, flips, and turns with pentominoes

1. There are four pairs of pentomino figures below. In each case, the top figure was transformed into the bottom figure by a flip or a turn. Describe what was done to the top figure to create the bottom figure. Use your pentomino pieces to check your answer.

a.   b.   c.   d.

2. Describe what was done to the top figure to create the bottom figure.

a.   b.   c.   d.

3. Use the grids on page 243. Make your own pentomino figures on the top grid. Then transform the figure by making any combination of flips or turns. Record the result on the bottom grid. Describe what you did. Give your problems to a friend and see whether your friend can guess your transformations.

Copyright © Houghton Mifflin Company. All rights reserved.

Figures for EXPLORATION 9.8, PART 1 and PART 2

PART 1: Equilateral triangles

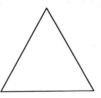 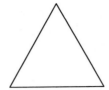 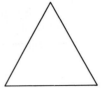

  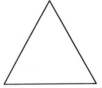

PART 2: Squares

Copyright © Houghton Mifflin Company. All rights reserved.

EXPLORATION 9.9 Tessellations

PART 1: Which figures tessellate?

Look at the following six pairs of figures. Work with several copies of the enlarged versions of the figures in order to determine which of the figures in each pair will tessellate: both, just one, or neither.

If both of the figures tessellate, make and justify a generalization: All figures that "look like" these two and have the following characteristics [describe the characteristics], will tessellate.

If just one of the figures in a pair tessellates, explain why the one does tessellate and why the other doesn't. That is, what characteristics or properties does the one figure have that the other doesn't have?

If neither of the figures tessellates, describe whether you believe that modifications of the figure could be made so that figures that "look like" these two might tessellate.

1.
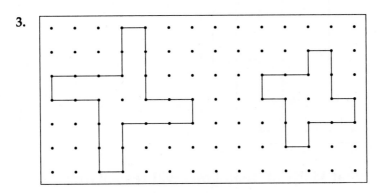

2.
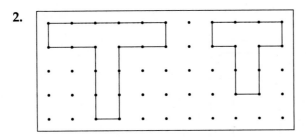

3.
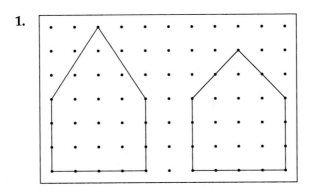

Copyright © Houghton Mifflin Company. All rights reserved.

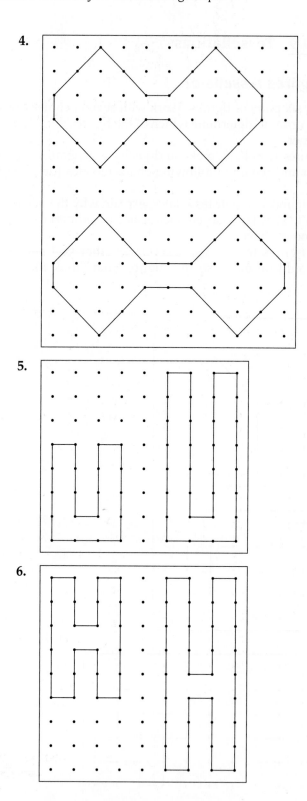

4.

5.

6.

PART 2: What kinds of pentagons will tessellate?

While the regular pentagon does not tessellate, some pentagons *do* tessellate. Under what conditions will a pentagon tessellate? Explore this question by making and testing several different kinds of pentagons.

   Tips: Fold a piece of paper in half, then in half again, and then in half again. Now if you draw a pentagon on the page and then cut out the pentagons (with heavy scissors),

Copyright © Houghton Mifflin Company. All rights reserved.

you will have eight copies of the pentagon. Alternatively, you can make a pentagon using a software program and cut and paste many copies of the pentagon.

Prepare a report that includes

- Your conclusion(s) in the following form: A pentagon with these characteristics [describe the characteristics] will tessellate. If you find more than one family that tessellates, write the description of the characteristics of each family.
- Your justification of your conclusion(s).
- A brief summary of your solution path: How did you come to your conclusion(s)? This will include your "failures" as well as your "successes."

PART 3: What kinds of arrows will tessellate?

Believe it or not, there are many kinds of arrows that will tessellate. In this exploration, your challenge is to determine the characteristics that are necessary in order for an arrow to tessellate. For example, does it need to be symmetric? Do the sides of the shaft need to be parallel? Can the tip be skinny? Make and test several different kinds of arrows, such as those shown below, either on blank paper or on Geoboard Dot Paper.

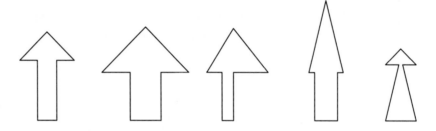

Prepare a report that includes

- Your conclusion(s) in the following form: An arrow with these characteristics [describe the characteristics] will tessellate. If you find more than one family that tessellates, write the description of the characteristics of each family.
- Your justification of your conclusion(s).
- A brief summary of your solution path: How did you come to your conclusion(s)? This will include your "failures" as well as your "successes."

PART 4: Semiregular tessellations

Now let us expand our discussion to combinations of figures that tessellate. A *semiregular tessellation* occurs when two or more regular polygons tessellate and every vertex point is congruent to every other vertex point.

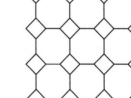

1. a. Using your understanding of congruent, try to define *congruent vertex point*.
    b. Compare definitions with your partner(s). Modify your definition, if needed, as a result of the discussion.

2. Cut out the figures on the Regular Polygons sheet at the end of the book. Use them to explore combinations of simple geometric figures to determine other semiregular tessellations.

    a. Sketch your "successes" and "failures" and record your reflections after each success or failure.

Copyright © Houghton Mifflin Company. All rights reserved.

   b.   How many semiregular tessellations do you think there are—5, 10, 20, 50, 100, thousands, an infinite number? Explain your reasoning.

  3. In your explorations, you may have found some *demiregular tessellations*, which occur when two or more regular polygons tessellate but not every vertex point is congruent to every other vertex point. Show any demiregular tessellations that you have discovered. Explain why they are not semiregular tessellations.

PART 5: Escher-like tessellations

Using the language of translations, reflections, and rotations, we are now in a position to understand how the artist M. C. Escher made his tessellations (see, for example, Figure 9.19 on page 593 of the text). Escher began with a polygon that would tessellate and then transformed that polygon. We can use the analogy of chess to understand how he did this: There are legal moves and illegal moves.

  1. *Translations* One legal move is to modify one side of the polygon and then translate that modification to the opposite side. For example:

Begin with a square (Step 1).

Modify one side (Step 2).

Translate the modified side to the opposite side (Step 3).

In this case, the translation is a vertical slide.

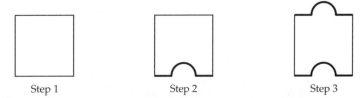

       Step 1   Step 2   Step 3

   a.   Begin with a square, and experiment with translations. Show your work and reasoning so that your instructor can follow your thinking. Make notes of your observations as they happen: patterns you see, conjectures you decide to pursue, questions you have.

   b.   Summarize these observations, patterns, and hypotheses about translations and tessellations.

  2. *Rotations* Another legal move is to rotate part of a shape. The center of rotation—that is, the "hinge" about which the part rotates—can be a vertex or a midpoint of one of the sides. Here is an example:

Begin with an equilateral triangle (Step 1).

Modify one side (Step 2).

Rotate that side 60 degrees counterclockwise about the bottom left vertex of the triangle (Step 3).

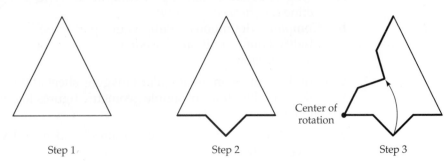

Center of rotation

       Step 1   Step 2   Step 3

Copyright © Houghton Mifflin Company. All rights reserved.

Quilt Blocks for EXPLORATION 9.10, PART 3

Baby Blocks

Broken Windows

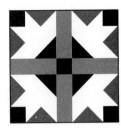

Cross Roads

Does and Darts

Drunkard's Path

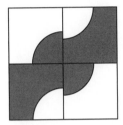

Flower Basket

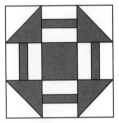

Hole in the Barn Door

Kansas Sunrise

Le Moyne Star

May Basket

Saw Tooth Star

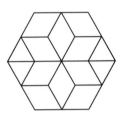

Star

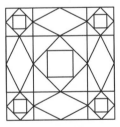

Storm at Sea

Snail's Trail

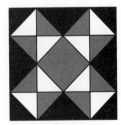

Old Tippecanoe

Tulip Basket

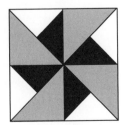

Windmill

Yankee Puzzle

Copyright © Houghton Mifflin Company. All rights reserved.

PART 4: Closest relative

Look at the quilt blocks below. Find the "closest relative" to Indian Star. Justify your choice. There is no "right" answer. The quality of your answer depends on your justification. This task involves describing the commonalities that you find that make the two patterns so similar. Note that this is a communication task and a reasoning task. How well you do depends partly on applying problem-solving tools and making connections.

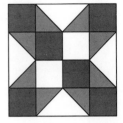

Indian Star

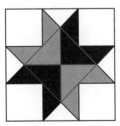

25-Patch Star

African Safari

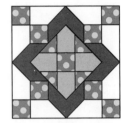

Grand Prix

Hearth and Home

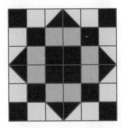

Oregon Trail

Prairies 9 Patch

Right Hand of Friendship

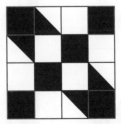

Road to Oklahoma

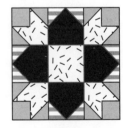

Weathervane

PART 5: Using transformations to construct a quilt pattern

Your instructor will ask you to construct a quilt pattern by beginning with the smallest possible piece(s) and then using slides, flips, and turns to generate the pattern. The example below illustrates one solution path for Dutch Man's Puzzle on p. 284.

Begin with the square shown in Step 1. Reflect this square through its right side (Step 2).

Translate this rectangle up one unit (Step 3).

Rotate the figure from Step 3 counterclockwise through a 1/4 turn (90 degrees); the center of rotation is the top left corner of the figure. The result is shown in Step 4.

Copyright © Houghton Mifflin Company. All rights reserved.

Rotate the figure from Step 4 counterclockwise through a 1/2 turn (180 degrees); the center of rotation is the midpoint of the left side of the figure. The result is shown in Step 5.

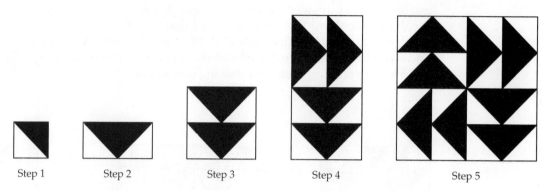

| Step 1 | Step 2 | Step 3 | Step 4 | Step 5 |

PART 6: Similarities and differences

For each pair of quilt patterns below, describe how the two patterns are alike and how they are different.

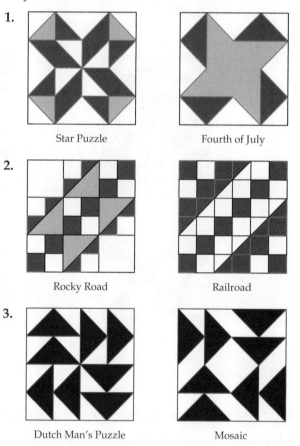

1.

Star Puzzle   Fourth of July

2.

Rocky Road   Railroad

3.

Dutch Man's Puzzle   Mosaic

Copyright © Houghton Mifflin Company. All rights reserved.

SECTION **9.3** Exploring Similarity

EXPLORATION 9.11 Similarity with Pattern Blocks

In Chapter 5, we used Pattern Blocks to explore fraction concepts. These blocks also let us see geometry "in action."

PART 1: Triangles

1. Take the small green triangle. Using the Pattern Blocks, make a bigger triangle that is similar in shape to this triangle. Explain why you believe this triangle is similar to the original triangle.

2. Make another triangle that is similar to the original triangle. Explain why you believe this triangle is similar to the original triangle.

3. Record other observations: patterns, conjectures, and questions. For example, are there patterns in the lengths of the sides? In how large the figures are? In how you actually make larger figures?

PART 2: Squares

1. Take the small orange square. Using the Pattern Blocks, make a bigger square that is similar in shape to this square. Explain why you believe this square is similar to the original square.

2. Make another square that is similar to the original square. Explain why you believe this square is similar to the original square.

3. Record other observations: patterns, conjectures, and questions. For example, are there patterns in the lengths of the sides? In how large the figures are? In how you actually make larger figures?

PART 3: Parallelograms

As it turns out, all equilateral triangles are similar to one another, and all squares are similar to one another. This is not true for all parallelograms!

1. Take the small blue parallelogram. Using the Pattern Blocks, make a bigger parallelogram that is similar in shape to this parallelogram. Explain why you believe this parallelogram is similar to the original parallelogram.

2. Make another parallelogram that is similar to the original parallelogram. Explain why you believe this parallelogram is similar to the original parallelogram.

3. a. Write your first draft of a definition of *similar* with respect to parallelograms. Your definition should be more precise than "same shape, possibly different size."

   b. Compare your definition with that of your partner(s). After the discussion, modify your definition, if needed.

4. Record other observations: patterns, conjectures, and questions. For example, are there patterns in the lengths of the sides? In how large the figures are? In how you actually make larger figures, and so on?

Copyright © Houghton Mifflin Company. All rights reserved.

PART 4: Trapezoids

1. Take the small red trapezoid. Using the Pattern Blocks, make a bigger trapezoid that is similar in shape to this trapezoid. Explain why you believe this trapezoid is similar to the original trapezoid.

2. Make another trapezoid that is similar to the original trapezoid. Explain why you believe this trapezoid is similar to the original trapezoid.

3. Record other observations: patterns, conjectures, and questions. For example, are there patterns in the lengths of the sides? In how large the figures are? In how you actually make larger figures, and so on?

4. Which of the following trapezoids is similar to the small red trapezoid? Justify your reasoning.

5. **a.** Write your first draft of a definition of *similar* that would be true for any geometric figure. Be precise, as noted in Part 3.
   b. Compare your definition with that of your partner(s). After the discussion, modify your definition, if needed.

Extension

There is a pattern concerning the area of consecutive similar pattern blocks that is true for all of the blocks: The ratio of the area of the next bigger similar figure to the original is 4:1. That is, if we count the area of the green triangle as 1 unit, the next bigger equilateral triangle has an area of 4 units. The blue parallelogram has an area of 2 units, and the next bigger blue parallelogram has an area of 8 units. Thus the ratio of their areas is 4:1. This is also true for the orange square, the red trapezoid, and the yellow hexagon.

Do you think this is true only in these cases, or will it be true for composite figures too? For example, will the area of the next bigger figure similar to this one be 4 times as great as the area of this one?

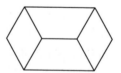

Copyright © Houghton Mifflin Company. All rights reserved.

EXPLORATION 9.12 Similar Figures

PART 1: Similar triangles

In high school geometry, we discovered that we did not have to show that all six pairs of corresponding angles and sides were congruent in order to know that two triangles were congruent. We learned about SSS, SAS, ASA, AAS, and HL. What might we need to show in order to know that two triangles are similar? At this point, in order to know that the two triangles are similar, we need to measure the lengths of all six sides and all six angles, and then we have to determine three ratios. If all the ratios are equal and all three pairs of angles are equal, then we know the two triangles are similar.

$$\frac{AB}{PQ} = \frac{AC}{PR} = \frac{BC}{QR} \quad \text{and} \quad m \angle A = m \angle P, m \angle B = m \angle Q, m \angle C = m \angle R$$

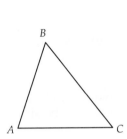 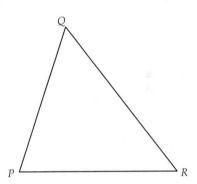

The question is: Do we have to know all the ratios to determine whether two triangles are similar? What combinations of sides and angles will be sufficient?

1. Write your initial hypothesis and reasoning about *one* combination that is sufficient to prove that two triangles are similar.

2. Discuss your ideas with your partner(s). Take one hypothesis; try to convince yourself that it is true, and try to make triangles that are not similar but that satisfy your hypothesis.

3. Present your hypotheses to another group. Note any comments or suggestions made by the other group.

4. Work through Part 1 repeatedly (as time permits) in order to convince yourself of as many similarity combinations as you can.

Copyright © Houghton Mifflin Company. All rights reserved.

PART 2: Similar quadrilaterals

The figure below shows a pair of similar rectangles and a pair of similar trapezoids. At this stage of our explorations, to confirm that they are indeed similar, we would have to measure all eight sides and all eight angles and verify that the lengths of corresponding pairs of sides had the same ratio and that corresponding angles were congruent. The question is: Do we have to know all eight pieces of information? What combinations of sides and angles will be sufficient?

1. *Rectangles*

    a. Write your initial hypotheses and reasoning for rectangles.
    b. Discuss your ideas with your partner(s). Take one of your hypotheses; try to convince yourself that it is true, and try to make rectangles that are not similar but that satisfy your hypothesis.
    c. Present your hypotheses to another group. Note any comments or suggestions made by the other group.

2. *Trapezoids*   Repeat Step 1 for trapezoids.

Extensions

Come up with a conjecture that will work for all quadrilaterals. That is, if these conditions are met, then the two quadrilaterals will be similar. Justify your conjecture

Copyright © Houghton Mifflin Company. All rights reserved.

EXPLORATION 10.7   Exploring Area on Geoboards

Let us explore the concept of area and some area formulas using Geoboards.

PART 1: Area on the Geoboard

In our explorations on the Geoboard and on Geoboard Dot Paper, we will discuss the areas of figures in reference to the unit square—that is, the area enclosed by the smallest square you can make on your Geoboard.

1. a. Make as many "different" squares as you can on a 5 × 5 Geoboard or on the Geoboard Dot Paper supplied at the end of the book.
    b. Compare your results and your strategies with your partner(s).
    c. Determine the area of each of the squares.
    d. Challenge: Can you make a square with an area of 1 unit? 2 units? 3 units, and so on? Which squares can't be made? Can you explain why they are impossible?

2. a. Make as many "different" rectangles as you can on a 5 × 5 Geoboard or on the Geoboard Dot Paper provided by your instructor.
    b. Compare your results and your strategies with your partner(s).
    c. Determine the area of each of the rectangles.
    d. Challenge: Can you make a rectangle with an area of 1 unit? 2 units? 3 units, and so on? Which rectangles can't be made? Can you explain why they are impossible?

3. a. Determine the area enclosed by each of the figures on page 301. Briefly describe your solution path so that a reader could see how you determined the area.
    b. Compare your answers and your strategies with your partner(s).
    c. With your partner, make an unusual shape on the Geoboard. Determine the area independently, and then compare answers and solution strategies. Do this for several figures.
    d. Summarize and justify useful strategies that you developed. Imagine writing this and sending it to a friend in order to share your insights.

Copyright © Houghton Mifflin Company. All rights reserved.

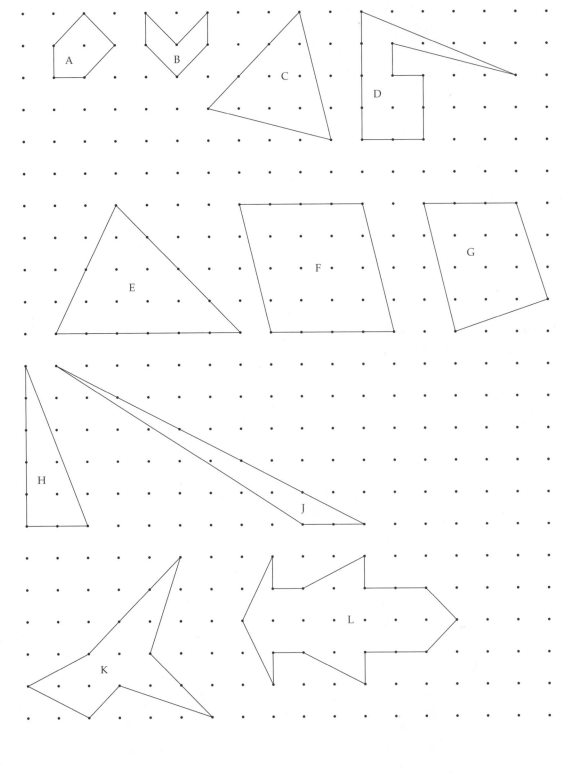

Copyright © Houghton Mifflin Company. All rights reserved.

EXPLORATION 10.12   Exploring Relationships Between Perimeter and Area

The relationship between perimeter and area is a rich field for exploration because there are many possible ways in which these two attributes can be related.

PART 1: Perimeters and areas on Geoboards

1. **a.** Suppose someone doubled the area of her garden. Does that mean that the length of the fence around the perimeter of her garden doubled also? What do you think?
   b. If we know that the area of one garden is 50 percent greater than the area of another garden, does this mean that the perimeter will be 50 percent greater? What do you think?
   c. Take some time to describe your present thoughts on the relationship between perimeter and area from this perspective.

   The relationship between area and perimeter is not a simple one, and we are going to investigate the relationship systematically, using a technique developed by mathematicians that helps us when relationships are complicated. We are going to keep one variable constant and look at what happens to the other variable.

2. **a.** On dot paper, make a number of polygons, each of which has an area of 15 units and in each of which all sides are either horizontal or vertical line segments. Determine the perimeter of each figure. In this case, we are holding area and certain attributes of shape constant and looking at how the perimeter changes.
   b. Look at those figures with the smallest perimeters and those with the greatest perimeters. Describe differences between the figures with smaller perimeters and the figures with greater perimeters.

3. **a.** Now make a number of polygons, each of which has a perimeter of 24 units and in each of which all sides are either horizontal or vertical line segments. Determine the area of each figure. In this case, we are holding perimeter and certain attributes of shape constant and looking at how the area changes.
   b. Look at those figures that have the smallest areas and those figures that have the largest areas. Describe differences between the figures with smaller areas and the figures with larger areas.

4. Finally, consider the original question: How are perimeter and area related? What do you believe now? Describe your present beliefs. If they are different from or more refined than your previous beliefs, describe the experiences, observations, or conversations that changed your beliefs.

PART 2: Changing dimensions

Let us explore the relationship between perimeter and area from another perspective.

1. What if you doubled the length of a rectangle but didn't change the width? The area would double, but what about the perimeter? What would be the effect on the perimeter? Explore this question with rectangles of different dimensions. Can you arrive at a statement that will be true for *all* rectangles? Write a report containing the following material:
   a. A brief summary of what you did.
   b. Your present description, in words and/or a formula, of the relationship between the original and the new perimeter.

2. Exchange descriptions with a partner. Provide feedback with respect to accuracy and clarity.

Copyright © Houghton Mifflin Company. All rights reserved.

PART 3: The banquet problem

Spaghetti and Meatballs for All! is a children's book written by Marilyn Burns to engage children in exploring the relationship between area and perimeter. Her story is a variation of a classic math problem called the banquet problem. Here we will explore this version: A family is having a get-together. They expect 18 people. They can rent card tables, each of which can seat 4 people. They have decided that they want to arrange the tables in the form of a rectangle. If they simply connect them in one long line, they will need 8 tables, because they can seat 8 people on each side and 1 person on each end. What are all the possibilities?

PART 4: Problems from the classroom

The following problems come from "Perimeter or Area: Which Measure Is It?" by Michaele Chappel and Denisse Thompson, which appears in the September 1999 issue of *Mathematics Teaching in the Middle School*, page 21. The authors report how students did on several questions.

1. Draw a figure whose perimeter is 24 units.

2. Here are two student solutions. Are they correct? Why or why not?

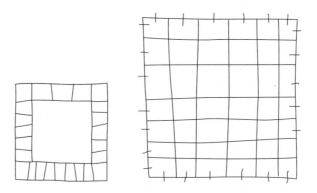

3. Write a realistic story problem in which you need to find the perimeter.

4. Here are three student stories. Discuss the stories in terms of how realistic they are and how valid they are.

    a. Alex has a table that is 100 centimeters long and 50 centimeters wide. What is the perimeter?

    b. Mary wants to border a cake with frosting. The cake is square, and one side is 6 inches. She uses a tablespoon for every inch around the cake. How much frosting does she need?

    c. Ashleigh is buying wallpaper for a room and knows that one wall is 7 feet across and that the room is a perfect square. Find the perimeter and tell how much wallpaper Ashleigh needs to buy.

5. This question was the hardest. Less than 8% of the students answered the problem correctly. Try it: Can two figures have the same area but different perimeters? Explain your answer.

Copyright © Houghton Mifflin Company. All rights reserved.

EXPLORATION 10.13   Functions, Geometric Figures, and Geoboards

Throughout this book, we have seen that there are many functional relationships in mathematics and in everyday life. In this exploration, we will discover relationships among three variables: the area of a figure, the number of pegs on the border of the figure, and the number of pegs in the interior of the figure.

PART 1: Exploring the relationships between pegs and right triangles

1. Use your Geoboards or the Geoboard Dot Paper supplied at the back of the book to fill in the top table on page 313. When you see a pattern, note it. Does that pattern lead to a hypothesis? For example, can you predict the area or number of border pegs or interior pegs in the next triangle?

2. When you feel that you are able to predict the area and the number of border and interior pegs when you are given only the dimensions of the triangle, state your hypothesis and how you came to discover it.

3. Predict the values for the three columns of the table for a 16 by 3 triangle and a 20 by 3 rectangle. Explain your predictions.

4. a. Use the top grid on page 315 to show the relationship between the length of the longer leg and the area of the triangle.
    b. What does this graph's being a straight line mean? Justify your response.

5. Use the bottom grid on page 315. In this step you are going to graph two different relationships on the same graph.

    a. First plot the points representing the relationship between the length of the longer leg and the number of border pegs. Then plot the points representing the relationship between the length of the longer leg and the number of interior pegs.
    b. What does neither of these graphs' being a straight line mean? Justify your response.
    c. Does it mean that they are not functions? Justify your response.
    d. Can you make use of these patterns to extend the graphs—that is, to predict the number of interior and border pegs when the longer base is 12, 13, 14, 15, or more units long?
    e. State your hypothesis now for predicting the number of interior and border pegs for any right triangle. Justify your hypothesis.

PART 2: Expanding our exploration to other figures

We are going to expand our exploration of relationships among border and interior pegs to include other figures. We are also going to change the focus slightly. Now, we are going to look for patterns so that we can predict the area of any geometric figure on the Geoboard from the number of border and interior pegs.

1. Make several geometric figures on your Geoboard or Geoboard Dot Paper. In each case, determine the area (A), the number of border pegs (B), and the number of interior pegs (I) and record those values in the bottom table on page 313. Do you notice any patterns that lead to hypotheses? If you do, record your hypothesis and how it came to be. Test your hypothesis. If it works, great. If not, back to the drawing board. If you do not see patterns that lead to hypotheses, move on to another geometric figure.

2. Now that you have determined the relationship among the three variables B, I, and A, look back and describe the moment of discovery.

Copyright © Houghton Mifflin Company. All rights reserved.

Tables for EXPLORATION 10.13, PART 1 and PART 2

PART 1: Exploring the relationships between pegs and triangles

1.

Dimensions of the right triangle	Area	Number of border pegs	Number of interior pegs	Insights/Observations
3 by 3	4.5 sq. units	9	1	
4 by 3				
5 by 3				
6 by 3				
7 by 3				
8 by 3				
9 by 3				
10 by 3				
11 by 3				

PART 2: Expanding our exploration to other figures

1.

Geometric figure	Area (A)	Number of border pegs (B)	Number of interior pegs (I)

Copyright © Houghton Mifflin Company. All rights reserved.

INDEX

Accuracy, 155, 293, 294, 295, 297, 308
Adding up algorithm, 53
Addition
　in Alphabitia, 51–52
　alternative algorithms, 52–53
　children's algorithms, 52
　of decimals, 121
　difference between addition and
　　multiplication, 64
　of fractions, 106
　of integers, 96
　mental, 51
　and number sense, 54
Additive comparisons, 132
Additive inverse, 97
African sand drawings, 90–91
Algebraic thinking, 24
　border growth pattern, 34–37
　connecting graphs and words, 29–30
　figurate numbers, 32–34
　magic squares, 13–14
　pattern block patio, 37
　pouring water into a bottle, 31
　relationships between variables, 27–28
Algorithms, alternative
　addition, 52–53
　division of fractions, 118
　division of whole numbers, 73
　multiplication, 67–68
　subtraction, 56–57
Algorithms, standard
　division, 74
　multiplication, 66
Alphabitia, 38–41
　addition in, 49–50
　subtraction in, 51
Ambiguity in measurements, 322
Angles, sums of, 210
　understanding, 202
Area
　of circle, 305
　cost of carpet and, 307
　formulas, 303–304
　on geoboards, 299–304
　irregular, 308
　meaning of, 298
　of parallelogram, 303
　perimeter and, 309–310
　surface area, 317
　of trapezoid, 304
　of triangle, 303
Area models
　of fractions, 104–106, 107–108
　of multiplication, 64–65
Arrows, tessellations of, 273
Assumptions, 8, 103

Attributes of polygons, 213–214
Average, 146–148

Base, of numeration system, 43–46
Base 2, 43–45
Base 6, 43–45
Base 10
　circle clocks, 60–61
　translating from, 45
　translating to, 44–45
Base 10 blocks
　for decimals, 120
Base 16, 44–45
Borrowing, 50
Boxes, volumes of, 318
Brahmagupta, 118
Bridges, 149
Bungee jump, 149

Carrying, 50
Census, 144
Center, measures of, 146–148
Chance, *see* Probability
Children's thinking, 4, 5, 16, 21, 25, 27, 34,
　　55, 56, 161, 176, 177, 201, 207, 215,
　　294, 295, 310
Circle
　area of, 305
Circle clocks, 60
Combinatorics, 170–173
Communication
　about geoboard figures, 175–176
　about three-dimensional figures, 219,
　　223–225
Comparison model, of subtraction, 97
Comparisons
　additive, 132
　multiplicative, 132
　of prices, 8, 135
Composite numbers, 80, 83, 89
Congruence
　with geoboards, 211
　with polyhedra, 219
　with tangrams, 211–212
Congruent vertex point, 273
Cross out algorithm, 52
Cross product algorithm, 68
Cross sections, 227
Cubes
　cross sections, 227
　nets for, 228
　from polyominoes, 228
　two-dimensional representations, 225
　volumes, 318
Cuisenaire rods, fractions and, 104–105
Cycles, 92–93

Cylinders
　cross sections, 227
　surface areas, 317
　volumes, 319

Darts, 9
Data
　analyzing with sets, 23
　collecting, 23, 27–28, 145–167, 293–295,
　　320–322
　distributions of, 145–167, 293–295,
　　310–322
　graphing, 145–157
　measures of center, 146–148
　survey, 159
Decimals, 120–130
　algorithms, 121
　with base 10 blocks, 120
　operations with, 127
　precision with, 315, 316, 343
　repeating decimals, 122–123
　target games with, 128
　see also Fractions
Decomposition
　of fractions, 100
　of whole numbers, 79
Definitions, 208–209
Demiregular tessellations, 274
Denominator, 102, 104, 110
Dependent variable
　growth pattern with, 35
　in lab investigations, 28
Diagonals of rectangle, 10
Dice games
　with whole numbers, 46
　operation sense in, 77
Discrete models, of fractions, 108
Distributions of data, 145–167
Divisibility, 79–93
Division
　alternative algorithms, 73
　computation, 70
　with fractions, 116–119
　of integers, 99
　mental, 71
　models of, 70, 99, 116
　and number sense, 75
　remainders in, 72
　scaffolding algorithm, 73
　standard algorithm, 74
　story problems with, 70
Dominoes, 187
　falling time, 27
Dot paper
　isometric, 224
　see also Geoboards
Duplation, Egyptian, 67

Copyright © Houghton Mifflin Company. All rights reserved.

Copyright © Houghton Mifflin Company. All rights reserved.

National Council of Teachers of
    Mathematics, *see* NCTM standards
Native American games, 171–173
NCTM standards, 1
  classifying and organizing information,
    23
  worthwhile mathematical tasks, 1
Negative numbers
  addition, 96
  division, 99
  multiplication, 98
  subtraction, 97
Nets, 228–230
Notation, translating into words, 33
Number line models, of fractions,
    104–105, 109
Number sense
  with decimals, 124–128
  with fractions, 113
  with whole numbers, 54, 58, 69, 75
  see also Operation sense
Number theory, 79–93
  African sand drawings, 90–91
  cycles, 92–93
  prime and composite numbers,
    80–89
Numeration systems
  Alphabitian, 38–40
  bases of, 43–45
Numerator, 102, 104, 110

Operation sense
  with decimals, 127
  with fractions, 114–117
  with whole numbers, 76–77
  see also Number sense

Paper folding, 253–260
Paper towels, measurements with, 321
Parallelograms
  areas, 303
  similar, 289
Pascal's triangle, 2–3, 17–20
Partial sums algorithm, 53
Partitioning model of division, 70
  with fractions, 116
  with integers, 99
Part-whole relationships, *see*
    Decomposition
Pattern Blocks
  fractions with, 104–105
  patio with, 37
  similarity with, 289–290
Patterns, 2
  and proof, 9–11
  and representations, 2–5
  African sand drawings, 90–91
  with angles, 202
  with different bases, 44
  in facial dimensions, 24
  in figurate numbers, 32–33
  functions and, 24
  in geometry, 221
  in magic squares, 13–14
  in multiplication, 6–7, 59–60
  with pattern blocks, 37, 289–290

of quilts, 278
  squares around border, 34–36
  in triangle puzzles, 15–16
Peanuts, 156
Pentagonal numbers, 33
Pentagons
  making with paper, 253
  tessellations of, 272–273
Pentominoes, 187–190, 242
  classifying, 188
  categorizing, 189–190
  challenges, 189
  defining, 187
  nets, 193
  similarity with, 245
  symmetry with, 245
  transformations with, 242
Percents, 137–142
Perimeter, 297, 309–310
Pi
  meaning of, 297
Piaget, Jean, 21
Place value
  in Alphabitia, 49–50
  in decimals, 120, 121
  in dice game, 46
  relative magnitue, 47
  with different bases, 43
Plane
  intersecting, 227
Polygons
  area versus perimeter, 309–310
  attributes, 213–214
  and relationship, 215–216
  sums of angles, 210
  symmetries of, 265
  tessellations of, 271–273
Polyhedra
  block buildings, 223–226
  comparing with polygons, 219
  congruence, 219–220
  cross sections, 227
  nets, 228–230
  regular, 222
  relationships among, 221
  sorting, 220
  surface areas, 317
  volumes, 318–319, 323
Polyominoes, 187–195
  classifying, 188
  defining, 187
  geometric transformations, 242
  similarity, 245
  symmetry, 245
  tessellations, 245
Populations
  data interpretation, 144
  density, 144
  sampling from, 167
Precision of mathematical language,
    208–209
Precision of measurements
  of area, 308
  assumptions and, 322
  in data projects, 158
  in lab investigations, 28, 308, 321, 322

linear, 293–295
  see also Estimation
Predictions (making), 2, 3, 5–7, 9–10,
    32–34, 80, 91, 121–123, 155, 228,
    246–247, 254, 261, 321
Prime factorization, 89
Prime numbers, 83
Prisms
  surface areas, 317
  vertices, faces, and edges, 219
  volumes, 318
Probability
  combinatorics, 170–173
  experimental, 160–167, 171
  fair games, 168
  sampling, 160–167
Problem-solving
  assumptions in, 8, 322
Proof
  and patterns, 9–11
  and geometry, 201, 210
Proportional reasoning, 131–142
  functions and, 136
  to interpret data, 133–134
  percents, 137–142
  pricing, 135
  rates, 133–134, 137–138, 142
  slopes of ramps, 132
Pulse, measuring, 156
Pyramids
  cross sections, 227
  surface areas, 318
  vertices, faces, and edges, 219
  volumes, 319

Quadrilaterals
  on geoboards, 160–177
  similar, 289–290, 292
  sum of angles, 210
Quadrinumbers, 136
Quilt patterns, 277–287, 306
  classifying, 279, 283
  closest relative, 283
  geometric transformations in, 283–284
  measurements of, 306
  reducing size of, 141
  symmetries of, 285

Ramps, slopes of, 132
Random sample, 159
Rates
  changes in, 133–134
  deaths from accidents, 133
  of population growth, 144
Rational numbers, *see* Decimals; Fractions
Ratios, *see* Proportional reasoning
Reaction time, 149
Real-life problems, 8
Rectangles
  areas, 310
  perimeters, 310
  similar, 292
Rectangular model
  of multiplication, 64–66
Rectangular prism, volume of, 318, 323
Reducing, 140

Copyright © Houghton Mifflin Company. All rights reserved.

Reflections (flips), 246–247
   on geoboards, 231–232
   with paper folding, 253–260
   with pentominoes, 242
   in quilt patterns, 283
   in tessellations, 274–276
Reflection symmetry
   on geoboard, 232
   with pentominoes, 245
   of polygons, 265
   in quilts, 285
   with tangrams, 245
Remainders, 72, 119
Repeated addition model, 116
   with fractions, 116
   with integers, 98
Repeated subtraction model, 116
   with fractions, 116
   with integers, 99
Representations, 2, 23, 83, 176–177, 207
Right bucket game, 124–126
Right triangles
   on geoboards, 311
Rotations (turns), 261
   on geoboards, 231–232
   with pentominoes, 242
   in quilt patterns, 283
   in tessellations, 274–276
Rotation symmetry
   of polygons, 265
   in quilts, 285

Sampling, 159–167
   random, 159
Scaffolding algorithm, 73
Scatterplot, 154
Scratch algorithm
   addition, 53
Semiregular tessellations, 273–274
Set model, of integer addition, 96
Sets
   children's thinking about, 21
   data about, 23
   Venn diagrams of, 22
Shapes, *see* Two-dimensional figures
Shooting hoops, 157
Similarity, 289–292
   with geoboards, 233
   with Pattern Blocks, 289–290
   with pentominoes, 245
   of quadrilaterals, 292
   of parallelograms, 289
   of rectangles, 292
   of squares, 289
   with tangrams, 241
   of trapezoids, 289
   of triangles, 289, 291
Simulations, 166
Slides, *see* Translations
Slopes
   of ramps, 132
Smallest amount game, 77
Snowflakes, paper, 260

Solids, *see* Three-dimensional figures
Spatial visualization, 219
Square inches, 298
Square numbers, 32–33
Squares
   areas, 299
   on geoboards, 299
   making with paper, 253
   similar, 289
Statistics, 143–159. *See also* Data
Story problems
   in Alphabitia, 49–50
   with division, 70
Subsets, *see* Sets
Subtraction
   and number sense, 58
   in Alphabitia, 50
   alternative algorithms, 56–57
   children's algorithms, 56
   of decimals, 120
   of fractions, 114–115
   of integers, 97
   mental, 55
Surface area, 317
Survey, 159
Symmetry
   with geoboards, 232–233
   groups, 266–268
   with pentominoes, 245
   of polygons, 265
   of quilt patterns, 285
   with tangrams, 241
Symmetry groups, 266–268

Tagging fish, 167
Take-away model, 97
Tangrams, 181–186
   classifying shapes, 185–186
   congruence with, 211
   puzzles, 181–182
   similarity with, 241
   symmetry with, 241
Target games
   with decimals, 128
   with whole numbers, 77
Taxman game, 80–81
Tessellations, 271–276
   of arrows, 273
   Escher-like, 274–276
   of pentagons, 271–272
   semiregular, 273–274
   with tetrominoes, 245
Tetrominoes, 187–189, 245
   tessellations with, 245
Theoretical probabilities, 160, 161, 171
Thickness, 295
Three-dimensional figures, 219–230
   classifying, 220
   communicating about, 219–220
   congruence with, 219
   cross sections, 227
   nets, 228–230
   relationships among, 221

surface areas, 317
   two-dimensional representations, 224
   volumes, 318–323
Transformations, *see* Geometric
   transformations
Translations (slides)
   on geoboards, 231–232
   with pentominoes, 242
   in tessellations, 274–276
Trapezoids
   areas, 304
   similar, 290, 292
Treviso algorithm
   for multiplication, 68
   for subtraction, 57
Triangles
   areas, 303–304
   as fraction models, 107
   with geoboards, 176–177
   similar, 289, 291
   sum of angles, 210
Triangular numbers, 32–33
Trinumbers, 32
Trominoes, 187
Turns, *see* Rotations
Two-dimensional figures
   classifying, 177, 185–186, 188, 189,
      215–216
   communicating about, 175–176, 187,
      204
   see also names of specific figures
Typical person, 145

Unit pricing, 135
Units
   in Alphabitia, 41
   in fraction problems, 103
   on geoboard, 299
   of measurement, 28, 298, 299

Valentine problem, 4
Variables
   dependent, 28, 35
   independent, 28, 35
   relationships between, 27–28
Variation, 143, 149–154, 156–157, 163, 320
Venn diagrams, 22, 215–216
Vertex point, congruent, 273
Vertices, of polyhedra, 219
Volume
   cereal box, 323
   cube, 318
   cylinder, 319
   irregular objects, 320
   pyramid, 319

What do you see?, 203, 205
Whirlybird, 152
Wholes, vs. units, 103
Word problems, *see* Story problems
Worthwhile mathematical tasks, 1

Zero, 38, 43
   in Alphabitia, 38

Copyright © Houghton Mifflin Company. All rights reserved.

CUTOUTS

Copyright © Houghton Mifflin Company. All rights reserved.

Base 10 Graph Paper

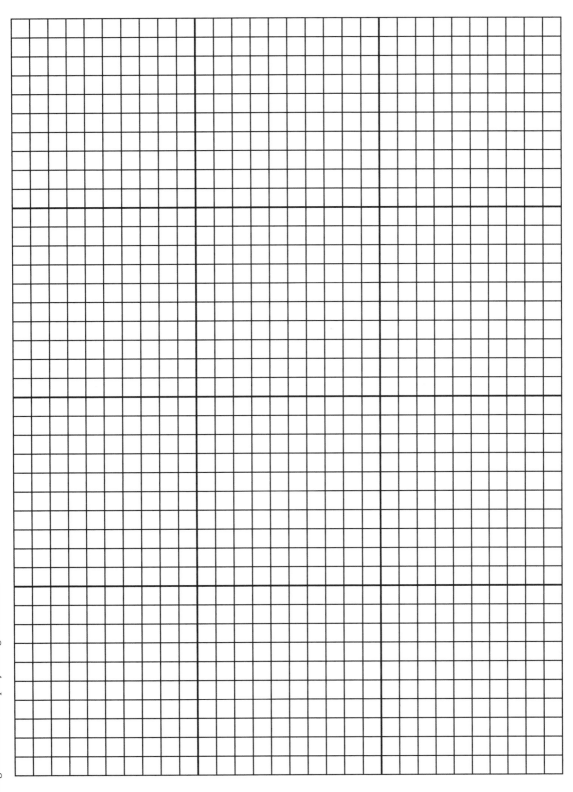

Copyright © Houghton Mifflin Company. All rights reserved.

Other Base Graph Paper

Copyright © Houghton Mifflin Company. All rights reserved.

Other Base Graph Paper

Copyright © Houghton Mifflin Company. All rights reserved.

Other Base Graph Paper

Copyright © Houghton Mifflin Company. All rights reserved.

Other Base Graph Paper

Copyright © Houghton Mifflin Company. All rights reserved.

Polyomino Grid Paper

Copyright © Houghton Mifflin Company. All rights reserved.

Polyomino Grid Paper

Copyright © Houghton Mifflin Company. All rights reserved.

Tangram Template

Copyright © Houghton Mifflin Company. All rights reserved.

Regular Polygons

Copyright © Houghton Mifflin Company. All rights reserved.

Exploring the Area of a Circle

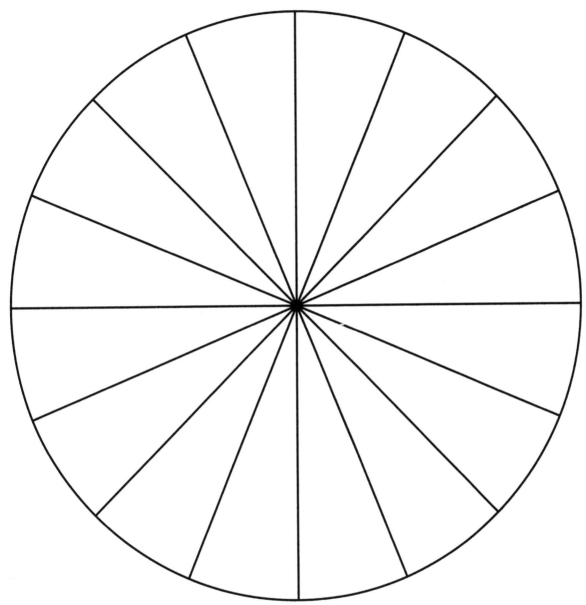

Copyright © Houghton Mifflin Company. All rights reserved.